SpringerBriefs in Applied Sciences and Technology

Manufacturing and Surface Engineering

Series Editor

Joao Paulo Davim, Department of Mechanical Engineering, University of Aveiro, Aveiro, Portugal

This series fosters information exchange and discussion on all aspects of manufacturing and surface engineering for modern industry. This series focuses on manufacturing with emphasis in machining and forming technologies, including traditional machining (turning, milling, drilling, etc.), non-traditional machining (EDM, USM, LAM, etc.), abrasive machining, hard part machining, high speed machining, high efficiency machining, micromachining, internet-based machining, metal casting, joining, powder metallurgy, extrusion, forging, rolling, drawing, sheet metal forming, microforming, hydroforming, thermoforming, incremental forming, plastics/composites processing, ceramic processing, hybrid processes (thermal, plasma, chemical and electrical energy assisted methods), etc. The manufacturability of all materials will be considered, including metals, polymers, ceramics, composites, biomaterials, nanomaterials, etc. The series covers the full range of surface engineering aspects such as surface metrology, surface integrity, contact mechanics, friction and wear, lubrication and lubricants, coatings an surface treatments, multiscale tribology including biomedical systems and manufacturing processes. Moreover, the series covers the computational methods and optimization techniques applied in manufacturing and surface engineering. Contributions to this book series are welcome on all subjects of manufacturing and surface engineering. Especially welcome are books that pioneer new research directions, raise new questions and new possibilities, or examine old problems from a new angle. To submit a proposal or request further information, please contact Dr. Mayra Castro, Publishing Editor Applied Sciences, via mayra.castro@springer.com or Professor J. Paulo Davim, Book Series Editor, via pdavim@ua.pt.

Panagiotis Kyratsis · Valerio Perna ·
João Paulo Davim

AI Models for Design and Digital Prototyping

Springer

Panagiotis Kyratsis
Department of Product and Systems Design Engineering
University of Western Macedonia
Kozani, Greece

Valerio Perna
Department of Architecture
EPOKA University
Tirana, Albania

João Paulo Davim
Department of Mechanical Engineering
University of Aveiro
Aveiro, Portugal

ISSN 2191-530X ISSN 2191-5318 (electronic)
SpringerBriefs in Applied Sciences and Technology
ISSN 2365-8223 ISSN 2365-8231 (electronic)
Manufacturing and Surface Engineering
ISBN 978-3-032-35143-2 ISBN 978-3-032-35144-9 (eBook)
https://doi.org/10.1007/978-3-032-35144-9

This Springer imprint is published by the registered company Springer Nature Switzerland AG
The registered company address is: Gewerbestrasse 11, 6330 Cham, Switzerland

Preface

The intersection of artificial intelligence (AI), design practice, and digital fabrication is undoubtedly one of the most promising and challenging fields of research for professionals and academics in the disciplines of design and architecture. Neural network architectures, LLMs, diffusion models, etc. have already shown an amazing ability to understand the space and generate novel geometries, synthesize spatial layouts, foresee and assess structural performances, speed up and optimize digital fabrication parameters and procedures, and understand and translate natural language descriptions into three-dimensional entities and forms. Such capabilities, besides the astonishing computational power that they embed, testify a qualitative shift from basic automation to genuine generation that contains in itself major implications for design educators, and practitioners are still trying to fully understand.

With a field of resonance intersecting different design scales—from the object to the building, and from the building to the city—AI is completely subverting the traditional ways in which design practices and production are conceptualized and actualized.

This book aims to provide a comprehensive exploration of AI-based design tools covering different design scales and approaches, and, at the same time, to serve both as an educational resource and a practical guide for designers and architects. Using a multi-faceted approach, the text offers valuable insights into the tools, methodologies, and best practices that characterize current existing AI-based workflows.

Structured into six chapters, it delves into topics such as the intellectual and historical path from early rule-based expert systems to contemporary generative architectures; the structure and entity of design problems at different scale at the emergence of new AI technologies; the inseparability of structure and aesthetic form; and the AI integration with CAD/CAM ecosystems and additive manufacturing pipelines. Furthermore, the Five-Phase AI-to-Fabrication Design Pipeline presented in last chapter represents the authors' original methodological contribution which aims not to focus on a single AI model or fabrication process but to be a methodology—a structured series of human and AI tasks with clear validation requirements and transition criteria at each phase boundary, intended to be adaptable across the variety of design scales, AI model families, and fabrication methods covered in the precedent chapters.

Finally, this volume aims to contribute towards a deeper understanding of the implication deriving from a conscious implementation of AI technologies in design practices, but also to inspire the readers to explore new possibilities and directions within this fascinating intersection between two different intelligences: the human and the artificial one.

Kozani, Greece — Prof. Dr. Panagiotis Kyratsis
Tirana, Albania — Assoc. Prof. Dr. Valerio Perna
Aveiro, Portugal — Prof. Dr. João Paulo Davim

Competing Interests The authors have no competing interests to declare that are relevant to the content of this manuscript.

Contents

1 A Review on AI Models for Design and Digital Prototyping 1
1.1 Introduction 1
1.2 Historical Trajectory: From Expert Systems to Generative Architectures 3
1.2.1 Early Computational Design and Expert Systems 3
1.2.2 The Machine Learning Transition 4
1.2.3 Transformer Architectures and the Foundation Model Era 5
1.3 AI Model Families Relevant to Design and Digital Prototyping 6
1.3.1 Convolutional Neural Networks 6
1.3.2 Generative Adversarial Networks 7
1.3.3 Variational Autoencoders 7
1.3.4 Transformer Based Architectures for Design 8
1.3.5 Diffusion Models 9
1.4 Three Modalities of AI in Design Practice 9
1.4.1 Optimization 10
1.4.2 Augmentation 10
1.4.3 Generation 11
1.5 Integration with CAD/CAM Ecosystems and Additive Manufacturing Pipelines 12
1.5.1 AI Integration Within CAD Platforms 12
1.5.2 AI in CAM and Process Planning 13
1.5.3 AI and Additive Manufacturing 13
1.6 A Taxonomy of AI Models by Design Task Type 14
1.7 Benefits, Challenges, and Open Research Questions 15
1.7.1 Demonstrated Benefits 15
1.7.2 Principal Challenges 15
1.8 Conclusion 16
References 17

2 The SMALL(S): AI for Product and Object Design ... 21
2.1 Introduction ... 21
2.2 The Object Scale: Characteristics and Design Challenges ... 23
2.2.1 The Inseparability of Structure and Aesthetic Form ... 23
2.2.2 Material as Form: The Object's Expressive Dimension ... 23
2.2.3 The Domestic Scale and the Human Body ... 24
2.3 AI-Driven Structural Performance Optimization of Designed Objects ... 24
2.3.1 Topology Optimization for Object Structures ... 25
2.3.2 Surrogate Models and Real-Time Structural Feedback ... 25
2.4 AI-Driven Form Generation and Aesthetic Exploration ... 26
2.4.1 Learning Object Typologies: GANs and the Design Corpus ... 26
2.4.2 Navigating Design Space: Variational Autoencoders and Morphological Exploration ... 27
2.4.3 Diffusion Models for Object Concept Visualization ... 28
2.5 Integrating Structural and Aesthetic AI: Compound Object Design ... 28
2.5.1 Sequential Multi-Modality Workflows ... 29
2.5.2 Performance-Informed Generative Design ... 30
2.6 AI-Enabled Object Customization and Mass Personalization ... 30
2.6.1 Parametric Design Systems and AI Interfaces ... 31
2.6.2 Anthropometric Customization: Objects Shaped by Bodies ... 32
2.6.3 Generative Design for Product Family Coherence ... 32
2.7 From AI-Generated Form to the Fabricated Object ... 33
2.7.1 Object Geometry and Fabrication Process Compatibility ... 33
2.7.2 Representation Conversion at the Object Scale ... 33
2.7.3 Craft, Finishing, and the Limits of Digital Fabrication ... 34
2.8 Benefits, Limitations, and Open Research Directions ... 34
2.8.1 Demonstrated Benefits ... 34
2.8.2 Principal Limitations ... 35
2.8.3 Open Research Directions ... 35
2.9 Conclusion ... 36
References ... 37

3 The MEDIUM(S): Diffusion Models and GAN for Building Scales and Single Artefacts ... 39
3.1 Introduction ... 40
3.2 The Architectural Scale: Characteristics and Design Challenges ... 41
3.2.1 Openness, Multi-Constraint, and the Problem of the Brief ... 41

3.2.2 Context, Site, and the Situated Character of Architecture ... 42
3.2.3 Program, Typology, and Spatial Logic ... 43
3.3 Facade Generation and Architectural Style Synthesis ... 43
3.3.1 Convolutional Approaches to Facade Analysis and Generation ... 44
3.3.2 Diffusion Models for Architectural Visualization ... 45
3.4 Floor Plan Generation and Spatial Layout ... 45
3.4.1 Image-Based GAN Approaches: The ArchiGAN Pipeline ... 46
3.4.2 Graph-Conditioned Generation: The House-GAN Approach ... 47
3.5 Structural Form-Finding at the Building Scale ... 47
3.5.1 AI-Assisted Structural Topology at the Building Scale ... 48
3.5.2 Machine Learning for Structural Performance Prediction ... 48
3.6 Creativity, Authorship, and Aesthetic Novelty in AI-Generated Architecture ... 49
3.6.1 Boden's Creativity Typology and AI-Generated Architecture ... 49
3.6.2 The Deep House: Estrangement as Design Method ... 50
3.6.3 Authorship, Attribution, and the Ethics of AI-Assisted Design ... 51
3.7 AI Integration with Building Information Modelling ... 52
3.7.1 The BIM Representation Gap ... 52
3.7.2 AI-Native Architectural Modelling Approaches ... 52
3.8 Benefits, Limitations, and Open Research Directions ... 53
3.8.1 Demonstrated Benefits ... 53
3.8.2 Principal Limitations ... 54
3.8.3 Open Research Directions ... 54
3.9 Conclusions ... 55
References ... 56

4 The LARGE(S): Defining the Urban Scale ... 59
4.1 Introduction ... 60
4.2 The Urban Scale: Characteristics and Design Challenges ... 61
4.2.1 Emergence, Self-organization, and the Limits of Design ... 62
4.2.2 Scale, Complexity, and Data ... 62
4.2.3 Time, Change, and the Urban Palimpsest ... 63
4.3 Urban Morphology Analysis and Figure-Ground Generation ... 64
4.3.1 Deep Learning for Urban Morphological Classification ... 64
4.3.2 Generative Models for Urban Figure-Ground ... 65

4.4 Urban Street Networks, Mobility, and Graph-Based AI 65
4.4.1 Graph Neural Networks for Street Network Analysis 66
4.4.2 Urban Scaling Laws and AI Urban Modelling 66
4.5 Generative Approaches to Urban Layout and Master Planning 67
4.5.1 From Building-Scale Generation to Urban-Scale Layout 68
4.5.2 Multi-Objective Urban Layout Optimization 68
4.5.3 Procedural Generation and Urban Growth Simulation 69
4.6 AI as Speculative Instrument: Urban Fictions and the Reimagination of the City 70
4.6.1 The Lineament and the Algorithm: Del Campo's Urban Fictions 70
4.6.2 The Nolli Plan as Training Data: Urban Memory and Machine Vision 71
4.6.3 Speculative AI Urbanism and the Discipline of Architecture 72
4.7 Ethical and Political Dimensions of AI Agency in Urban Design 73
4.7.1 Representation, Bias, and the Training Data Problem 73
4.7.2 Participation, Agency, and the Democratic City 74
4.7.3 AI and the Right to the City 75
4.8 Benefits, Limitations, and Open Research Directions 75
4.8.1 Demonstrated Benefits 75
4.8.2 Principal Limitations 76
4.8.3 Open Research Directions 77
4.9 Conclusions 77
References 78

5 Platforms and Ecosystems for AI-Driven Designing and Prototyping 81
5.1 Introduction 82
5.2 The Design-To-Fabrication Technology Stack: A Conceptual Framework 83
5.2.1 Layers of the Design-To-Fabrication Stack 84
5.2.2 Bottlenecks and the Representation Gap 84
5.3 Parametric Design Environments and AI Integration 85
5.3.1 Rhinoceros and Grasshopper: The Research Platform of Record 85
5.3.2 Industrial CAD Platforms and Generative Design 87
5.3.3 Oxman's Material-Based Design Computation: A Foundational Framework 87
5.4 AI-Native Creative Platforms and Their Relationship to Design Practice 88
5.4.1 Image Diffusion Platforms in Design Practice 89

5.4.2 Large Language Models as Design Assistants 90
5.4.3 Three-Dimensional Generation Platforms 91
5.5 Fabrication-Integrated AI Pipelines 92
5.5.1 AI-Assisted Process Planning and Toolpath Generation 92
5.5.2 AI for Additive Manufacturing Process Control 93
5.5.3 Robotic Fabrication and AI-Controlled Assembly 93
5.6 API Ecosystems and the Orchestration of Multi-Model AI Workflows 94
5.6.1 REST APIs and Cloud AI Services 94
5.6.2 Multi-Model Pipeline Orchestration 95
5.6.3 Open-Source Ecosystems and the Democratization of AI Design 96
5.7 Platform Dependency, Design Agency, and the Political Economy of AI Design Tools 96
5.7.1 Vendor Lock-In and the Autonomy of Design Practice 96
5.7.2 Data, Training, and the Ownership of Design Intelligence 97
5.8 Benefits, Limitations, and Open Research Directions 98
5.8.1 Demonstrated Benefits 98
5.8.2 Principal Limitations 98
5.8.3 Open Research Directions 99
5.9 Conclusions 99
References 100

6 From AI Generation to Digital Prototyping: A Methodological Framework 103
6.1 Introduction 104
6.2 The AI-To-Manufacture Gap: A Cross-Scale Synthesis 105
6.2.1 The Gap's Structural Origins 105
6.2.2 The Gap's Operative Manifestations 106
6.3 The Five-Phase AI-To-Fabrication Design Pipeline 107
6.3.1 Phase 1—Brief Definition 107
6.3.2 Phase 2—AI-Assisted Ideation 108
6.3.3 Phase 3—Geometric Reconstruction 108
6.3.4 Phase 4—Prototyping and Iteration 109
6.3.5 Phase 5—Test and Evaluation 110
6.4 Experimental Case Study: AI-Generated Backpack from Ideation to Fabrication 110
6.4.1 Phase 1—Design Brief and Constraint Specification 111
6.4.2 Phase 2—AI-Assisted Ideation and Concept Generation 111
6.4.3 Phase 3—Geometric Reconstruction and Mesh Validation 112

6.4.4 Phase 4—Fabrication and Process Documentation 114
6.4.5 Phase 5—Test and Evaluation 116
6.5 Framework Validation: Synthesis of Findings 117
6.6 Prospective Developments: Towards a Closed Design-To-Fabrication Loop 119
6.7 Conclusion 120
References 121

Chapter 1
A Review on AI Models for Design and Digital Prototyping

Abstract The intersection of artificial intelligence (AI), design practice, and digital fabrication is one of the most promising and challenging fields of research for professionals and academics. This chapter provides a structured review of different AI model families—spanning from convolutional neural networks (CNNs) to generative adversarial networks (GANs); from variational autoencoders (VAEs) to transformer-based architectures, and diffusion models—that are actually impacting the fields of design automation and digital prototyping. The review follows an intellectual and historical path that covers from early rule-based expert systems to contemporary generative architectures, and it discusses the functioning of the three main operative modalities in which AI can contribute in enhancing design practice (optimization, augmentation, and generation). Furthermore, it critically discusses the integration and diffusion of AI models within CAD/CAM ecosystems and additive manufacturing methodologies and production pipelines. The text is structured around a taxonomy of AI model families organized according to different design tasks, and it identifies a series of recognized benefits, limitations, and future challenges to be addressed. Moreover, the chapter serves as a theoretical and methodological ground to develop the scale-specific investigations presented in the other chapters of this book, covering from product design to architecture and the urban scale.

Keywords Artificial intelligence · Generative design · Diffusion models · Generative adversarial networks · Convolutional neural networks · Digital prototyping · Additive manufacturing · CAD/CAM · Design automation · Neural networks

1.1 Introduction

The act of design has always revolved around its capacity of being able not only to generate forms and shapes, but also to evaluate and refine them. For many centuries, this was considered an innate and peculiar ability of the human designer—mediated through history by use of different instruments, material knowledge, and the

P. Kyratsis et al., *AI Models for Design and Digital Prototyping*,
Manufacturing and Surface Engineering, https://doi.org/10.1007/978-3-032-35144-9_1

accumulation of disciplinary knowledge through experience and perseverance. In the 1960s, the emergence of Computer Aided Design (CAD) represented the first attempt at shifting this process towards a digital and computational substrate [1], allowing a significant amount of precision and standardization/reproducibility that human intellect and manual methods could not guarantee. Nevertheless, early CAD were no more than mere executors: they could not perform any action that was not specified by the human counterpart. Therefore, the only automation they could offer was the elimination/reduction of drafting labor.

A major shift arrived; the CAD systems were able to integrate programming features—namely CAD-based programming—and enable the parametric definition of design relationships, the automation of specific engineering tasks, and the use of Application Programming Interfaces (APIs) [2, 3] which offered the possibility to extract and subsequently manipulate geometric data. Such novelties made possible the transformation of CAD technologies into an active design environment, different from the original passive documentation medium that it used to be. Nevertheless, there was not even the most remote possibility that a system itself could propose, generate, or elaborate design solutions. The early scripting-based approach was still deterministic, where the designer was the one allowed to formulate rules and the software could at most perform some actions based on that. The rise of machine learning—with a fundamental role played by deep learning—completely transformed this picture. Neural network architectures, LLMs, diffusion models, etc. have already shown an amazing ability to understand the space and generate novel geometries [4], synthesize spatial layouts [5], foresee and assess structural performances [6], speed up and optimize digital fabrication parameters and procedures [7], and understand and translate natural language descriptions (namely prompts) into three-dimensional entities and forms [8]. Such capabilities, besides the astonishing computational power that they embed, testify a qualitative shift from basic automation to genuine generation that contains in itself major implications for design educators and which practitioners are still trying to fully understand.

This chapter has the objective to set a common intellectual ground for the further investigations contained in this volume. To do this, it relies on four principal contributions. First, it explores the major historical transitions from rule-based expert systems towards statistical learning and contemporary generative tools, with a specific focus on situation current capabilities inside their developmental context. Second, it analyses the most relevant AI model families that are reshaping the fields of design and digital prototyping—variational autoencoders (VAEs), generative adversarial networks (GAN), convolutional neural networks (CNNs), diffusion models, and transformer-based architectures—and, at the same time, it tries to identify their characteristics, strengths, but also current limitations. Third, it organizes AI models according their design task type through a structured taxonomy, to facilitate researchers and practitioners in choosing the best model to fit their needs. Fourth, it investigates the possible integration of such models in CAD/CAM environments and digital fabrication procedures, to address how AI-generated outputs can be easily translated in printable and manufacturable artefacts. The chapter ends with a structured synthesis of some identified challenges and discusses possible future research directions in the field.

1.2 Historical Trajectory: From Expert Systems to Generative Architectures

To understand the current abilities and behaviour that artificial intelligence (AI) models can show in design-based operations, it is needed to situate into a broader picture intersecting the fields of computational design and artificial intelligence from an intellectual perspective. During the last decades, the two-mentioned fields were developed largely in parallel until their recent convergence, when their peculiar characteristics found a fertile ground of intersection.

1.2.1 Early Computational Design and Expert Systems

The origins of computational design can be traced in the drafting automation systems researched and developed at MIT—and other institutions—during the 1960s: attempts that found their best outcome in Ivan Sutherland's landmark Sketchpad system (1963), which introduced interactive graphical computing and the concept of constraints as a mechanism for establishing and maintaining geometric relationships [1]. These almost rudimentary early CAD systems were followed, in the 1970s and 1980s, by the development of solid modelling—the representation of three-dimensional geometry through mathematic formulas thus as precise volumetric objects—and parametric modelling, which enabled dimensions and constraints to propagate changes automatically through a design model [9].

The first intersection of artificial intelligence with design during this period can be found in the so-called knowledge-based expert systems: basically, some rule-based programs that encoded domain expertise in the form of *if–then* production rules and applied it deterministically to design problems through inference engines. The main contribution to the field given by expert systems such as XCON (for computer configuration)—and various structural engineering advisory systems—was the demonstration that a specific domain knowledge (engineering, architecture, but also medicine for example) could be reduced and computationally formalized and then applied design problems defined by constraints [10]. Nevertheless, these first systems were intrinsically fragile: since they could not learn from experience, they could not even generalize beyond their encoded rule sets, and indeed required laborious and rigid manual knowledge acquisition by domain experts since deviations from the original matrix were not considered possible. For this reason, their usage to the field of design problems, known for its open-ended and creative thinking process, was severely limited and, above all, not considered productive to obtain meaningful outcomes.

It is in the 1990s that the design community saw the development of evolutionary and genetic algorithms as a qualitative-based different approach to enhance computational design exploration. Therefore, instead of mechanically encoding limited expert knowledge, these methods encoded design representations as variable parameters and applied automatic selection procedures based on defined fitness criteria: in this way, it became possible to generate a whole genealogy of candidate designs and iteratively refining them through processes analogous to biological evolution [11]. Such evolutionary approaches that demonstrated genuine exploratory capability—intended as the machine ability to discover non-intuitive configurations that a human designer might not have considered—were still computationally expensive, weak towards problems derived from wrong coding/formulation, and limited in the structural complexity of the design representations they could efficiently search.

1.2.2 The Machine Learning Transition

The transition *if–then* systems to a more structured computational design intelligence was possible thanks to three overlapping and parallel developments that happened in a time period spanning from the 2000s to early 2010s: (1) the accumulation of large digital design datasets (possible to already broad adoption of CAD software and tools, and the progressive digitalization of huge design archives); (2) the increase in available computational power allowed by the diffusion of graphics processing units (GPUs) for general matrix computation; (3) the theoretical and practical advances in deep neural network training, especially concerning the development of the capacity to effectively train neural networks faster and better [12].

In 2012 [13], ImageNet Large Scale Visual Recognition—in a few words the application of convolutional neural networks (CNNs) to image recognition tasks—showed not only that deep learning could match human performances but also exceed them in terms of complex pattern recognition if the dedicated neural network had been trained with a sufficient amount of data and could count adequate computational resources. On a design perspective, this achievement marked the beginning of a new era: considering that design outputs are commonly expressed through bidimensional representations and projections, the learning capacities of those models to learn from image data opened the possibility of training them through a large collection of design drawings, photographs, renderings, etc.

As a result, the following years saw a rapid development of deep learning architecture that was oriented to satisfy generative tasks: indeed, neural networks not anymore asked to classify existing artefacts but rather to generate novel outputs. The introduction of Generative Adversarial Networks (GANs) by Goodfellow et al. in [4] represented a conceptual breakthrough: for the first time, it was possible to train two networks in adversarial competition—on the one hand, a generator producing synthetic outputs from a given dataset; on the other, a discriminator attempting to distinguish them from real examples—that could produce high-quality synthetic images that could not be distinguished from their photographic original counterpart. Furthermore, Variational Autoencoders (VAEs), introduced by Kingma and Welling in the same year [14], provided a complementary approach: they encoded inputs into a probabilistic latent space while—at the same time—they were learning to decode samples from that space to turn them into plausible outputs, enabling structured interpolation and attribute manipulation in the latent representation.

1.2.3 Transformer Architectures and the Foundation Model Era

Originally developed for natural language processing purposes, the Transformer Architecture—introduced by Vaswani et al. in [15]—is considered as the most fundamental innovation of the decade in terms of architecture for AI in general, and even more when it comes to design AI. The Transformer's so-called self-attention mechanism, allowing each element into a sequence to attend to every other element with learnt weights, proved an exceptional capability for modelling long-range and interconnected dependencies in relation to sequential data. Grounded on this, the subsequent development of large-scale language models trained on web-scale text corpora—such GPT [16], BERT [17], and their successors—demonstrated that if models have sufficient scale, and they are trained on large and diverse datasets, they are able to exhibit more than the capacity they were explicitly trained for, as for example logical reasoning, code generation and optimization, and the ability to follow complex and structured natural language instructions.

Another fundamental developmental step was the possibility given to transformer architectures to proficiently implement multimodal data—e.g., text combined with images (CLIP [18]), with three-dimensional geometry, and with audio—that established the foundation for AI systems to translate natural language design descriptions into visual or geometric outputs. Models such as DALL·E™ [19], Stable Diffusion™ [20], and Midjourney™ are a concrete example of the possibility to generate detailed photorealistic and artistic images from natural language prompts and also the market-value that this possibility represents. As a results, the adoption of these capabilities in the fields of architectural visualization, product design concept generation, and urban design ideation has been rapid and consequential.

Nowadays, diffusion models are considered the dominant paradigm for generating high-quality (and high-fidelity) visual generative synthesis through the denoising of a random signal through a reversed diffusion process [21]. Furthermore, their ability to conditioning on auxiliary inputs (text prompts, sketch inputs, semantic maps) have made them the most chosen architecture for the global community of architects, designers, and three-dimensional artists. The following section explores the implications of these developments under a broad spectrum of design practices and procedures.

1.3 AI Model Families Relevant to Design and Digital Prototyping

Currently, there is a wide number of AI models that can be used in design and digital prototyping, and each one of those model families shows different and peculiar strengths in relation to the different tasks that are at the core of design practice. This section aims to offer a systematic analysis of the principal model families, with a focus on their operational principles, characteristic capabilities, and already known limitations for what concerns design contexts and tasks.

1.3.1 Convolutional Neural Networks

Convolutional Neural Networks (CNNs) are deep learning architectures specifically designed for the analysis of grid-structured data or, in a few words, bidimensional images. Their main characteristic is precisely the convolutional layer that has the function to apply learnt filters across the spatial extent of the input that allows the detection of local features at multiple scales. The succession of different convolutional layers guarantees a learning feature increasingly addressing abstract representations: for example, early layers can detect edges and local texture while deeper ones might be able to interpret and divide complex semantic patterns like specific object parts or situated spatial configurations [13].

In design contexts, CNNs have found a considerable application in the analysis and classification of design outputs: they can recognize and catalogue architectural styles starting from facade images [22]; they can classify product forms following aesthetic category; they can inspect images and detect structural defects in manufactured components; and they can perform a segmentation of building footprints and urban features starting from satellite and aerial imagery [23]. Even though they do not present generative capacities but rather classification and detection-based ones, they play acritical roles in design evaluation pipelines—enabling automated quality assessment, style consistency checking, and performance evaluation of large design datasets.

CNNs have also been adapted for what are called design-adjacent generative tasks through encoder-decoder architectures such as U-Net [24], which can preserve the spatial structure while at the same time perform transformation of an image into another. Some of the most common applications include the transition of architectural sketches into rendered images, the generation of floor plans from spatial constraint diagrams, and the conversion of semantic maps to photorealistic renderings [5]. Nevertheless, their dependence on fixed-resolution, grid-structured inputs, and the difficulty in modelling long-range spatial relationship is one of their main limitations for design generation, a limitation that was addressed by the development of transformer-based architectures.

1.3.2 Generative Adversarial Networks

Generative Adversarial Networks (GANs) functioning is rooted into the simultaneous training of two distinct neural networks that operate through an adversarial competition: a generator network G, with the task of mapping random noise vectors to synthetic data samples, and a discriminator network D, which is entitled to distinguish synthetic generated samples from real training data [4]. The training process is based upon a minimax game in which G wants to maximize the probability that D produces incorrect answers and fails in distinguishing its fake outputs from real dataset images, while D attempts to maximize its classification accuracy and discriminate G's outcomes. All these processes are needed to be sure that the generator could produce samples that are indistinguishable from the training set.

The introduction of conditional GANs (cGANs) [25], where both G and D are supported through auxiliary conditioning information, enhanced their working framework towards controlled generation implementing the possible specification of desired attributes such as building style, spatial program, and even just material palette. The implementation of the pix2pix framework [26] allowed conditional GANs to be used in image-to-image translation with design-relevant applications such as sketch-to-rendering, plan-to-elevation, and semantic-map-to-photorealistic conversion. A peculiar subfamily is the one represented by CycleGAN [27] that can grant unpaired image-to-image translation without requiring an input–output example or aligned training pairs, making it the perfect choice in case of scarce training data. In regards to architectural design, GANs have seen a broad application in the generation of floor plan layouts [5], building facade variations based on style, material, etc. [28], and the exploration of spatial configurations influenced by program requirements represented as a set of graph structures. In this perspective, House-GAN and its successors [5] are a relevant spin-off solution. Based on relational graphs embedding architectural programs—through which they condition the GAN—they showed an interesting capacity to produce spatially coherent floor plans that satisfy an early-stage architectural design exploration such as adjacency requirements. In terms of limitations, two major problems have been identified until now: model instability and the possibility for the generation to converge in a limited subset of the training distribution that could disorientate the discriminator and with the results not representing the full diversity expressed in the training dataset.

1.3.3 Variational Autoencoders

Variational Autoencoders (VAEs) address generative modelling from an information-theoretic perspective, using an encoder that maps inputs towards a probability distribution over a latent space and a decoder that maps latent samples back to the data space, and encourage the latent distribution to approximately reach a standard Gaussian [14]. This structured and continuous latent space from VAE training allows a smooth interpolation between several design variants and principled manipulation of design attributes.

In a more specific design context, VAEs have been applied to the learning of compact latent representations of three-dimensional shapes [29], enabling the generation of novel geometries by sampling or interpolating in the latent space. These capabilities have shown particular relevance in the field of product design exploration, where the ability to navigate a space of coherent and related form variations—a morphological design space—supports the exploration of design alternatives based on a common family of forms and shapes. VAEs have also been used for the latent representations of floor plan typologies [30], enabling designers to navigate between spatial configurations and explore the continuous territory between known plan types.

Their primary limitation—in comparison to GANs and diffusion models—is the blurriness that characterizes the generated outputs. This phenomenon is the consequence of the reconstruction loss, which tends to favour more average and smooth outputs over clearer and more specific ones. To address this limitation, Hybrid architectures combining VAE-style latent space learning with GAN-style adversarial training (VAE-GANs) have been developed to address this limitation.

1.3.4 Transformer-Based Architectures for Design

The recent years have seen a growing interest in the adaptation of transformer architectures for design tasks transforming this category in one of the most active areas in AI research. The transformer's self-attention mechanism, efficient when it comes to model relationships between all elements of an input sequence while respecting the weights of each component, has proven to offer satisfying in interpreting all the relational structures of spatial design problems: rooms in relation to corridors, facades in relation to plans, components in relation to assemblies [15].

Point-cloud transformers have demonstrated strong performance on three-dimensional shape analysis tasks, learning representations of unstructured point sets that support shape classification, part segmentation, and completion [31]. For design applications, this translates to the capacity to analyse and understand three-dimensional design geometry in its native, non-discretized form—without the information loss associated with voxelization or the topological assumptions of mesh-based representations. Autoregressive transformer models trained on sequences of design operations have demonstrated the capacity to generate CAD construction sequences—ordered series of geometric operations that, when executed, produce a valid three-dimensional model [32]. This capability is particularly significant for integration with CAD systems, as it produces design outputs in a format that CAD platforms can directly interpret and that designers can subsequently edit.

Large multimodal transformers, trained on paired text-image-geometry datasets, have enabled text-conditioned three-dimensional shape generation—the synthesis of three-dimensional geometry guided by natural language descriptions [8]. These capabilities are discussed further in the context of text-to-3D generation in Sect. 1.4.

1.3.5 Diffusion Models

Diffusion models are the dominant paradigm for high-fidelity generative synthesis across image, audio, video, and are continuously increasing their capacity in relation to three-dimensional data modalities. They operate by learning to reverse a gradual noising process: from a specific data sample, a forward diffusion process adds Gaussian noise progressively until the original data is reduced to pure noise; after that, a learnt denoising network is able to predict and subsequently remove this noise. Indeed, the process starts from a generated random noise, which is iteratively cleaned up to generate a synthetized outcome through a reverse process [21].

Contemporary latent diffusion models (LDMs) [20] are able to perform this diffusion process in a compressed latent space rather than in pixel space, thus maintaining the output quality while sensitively reducing the needed computational power, reducing computational requirements while maintaining output quality. LDMs use additional inputs—such as embedded texts, sketch and graphic inputs, and even semantic maps—to condition the denoising process and guarantee a fine-grained generation: the preferred style, content, or spatial configuration is indeed specified through natural language or a visual prompt/input. For these characteristics, diffusion models have become a primary choice in the fields of architecture and design, giving to these professionals the possibility to generate multiple elaborated visual concepts from textual or diagrammatic inputs [33]. An active area of research is the one interested in testing the capacities of diffusion models in managing three-dimensional data. Some of the most promising approaches include diffusion over point, diffusion over neural radiance field (NeRF) parameters, and the use of pre-trained image diffusion models to guide three-dimensional generation through iterative multi-view consistency optimization [34]. Nevertheless, even though three-dimensional diffusion models are still far from the needed spatial resolution and geometric precision to be fully integrating into manufacturing production pipelines, the continuous improvements testify that the distance between AI-generated conceptual geometry and fabrication-ready models is progressively being covered.

1.4 Three Modalities of AI in Design Practice

After analysing in the precedent section, the diverse existing AI model families, this chapter aims to draw a consistent analytical framework to understand how AI models engage with design practice. The proposal argues that current AI applications to design can be classified according to three operative modalities: optimization, augmentation, and generation. These modalities are not mutually exclusive—indeed, many systems combine elements of more than one—but they represent some distinct primary orientations that carry different implications affecting the role of the designer, the nature of the outputs produced from and with these tools, and the appropriate integration strategies into everyday design practices.

1.4.1 Optimization

The optimization modality challenges AI models to search within a defined design solution space for configurations that can maximize (or minimize) one or more declared performance metrics. In terms of design space, this is usually organized according to specific parameters from the design, architect, or engineer, where the AI model—whether a neural network trained to overcome computationally expensive simulation, a reinforced learning agent, or an evolutionary optimizer—explores this space of possible outcomes in a more efficient way than traditional methods.

One of the most interesting design applications in terms of optimization is surely topology optimization: namely the determination of an optimal distribution of material in a structural domain which is under load and/or boundary conditions [35]. Different from more traditional methods—such as SIMP or BESO, based on iterative finite elements—neural network models have the quality not only to perform complex calculations for optimized topologies but also to do at a fraction of the needed computational cost [36], while allowing real-time design feedback in early-stage explorations. Some of most common application of this surrogate models includes fluid dynamics for aerodynamic form optimization [37], daylight simulation concerning building envelope optimization, and thermal performance assessment for product enclosure design. The optimization modality shows a coherent continuity with the designer's traditional role: the AI searches for solutions that can satisfy the objectives and constraints specified by the human counterpart. The primary limitation in existing procedures lies in the impossibility for the generated outcomes to fall outside the predefined solution space: therefore, the model can produce online solutions that are valid within the parameterization previously set by the designer; thus it cannot generate anything outside of that domain.

1.4.2 Augmentation

In the augmentation modality, AI models support the designer's generative process through intelligent assistance, automated completion, or feedback. While the creative part is still in the hands of the designer, the AI system works as a helping system able to provide faster iterations, reduce frictions and errors, and extend the analytical capabilities of the designer.

Some of the most diffused applications in this category are the ones related with search and retrieval, where the AI tools can find relevant precedents or elements in a huge database through natural language or sketch queries [38]; produce automated drawing annotation and dimensioning and generate thorough technical documentation also for three-dimensional models [39]; provide real-time structural performance feedback and manage iterative performance assessment loops without any manual simulation setup; propose plausible completion for unfinished design parts using workflows similar to the common autocomplete function in text editing but enhancing this practices into spatial and geometric representations. In terms of diffusion and usage, the augmentation modality is currently the most mature and widely adopted protocol for a series of reasons: the technical demand is less impacting than full generation; the integration with already existing workflows is more straightforward; the level of preserved human agency is still balanced and somehow reassuring for both institutional clients and professionals. A major for this modality issue—especially for both educational actors and professionals—is the so-called skill atrophy, meaning the erosion of the designer's capacity due to the continuous delegation to AI system to perform difficult tasks.

1.4.3 Generation

In the generation modality, AI models produce design outputs—spatial configurations, three-dimensional geometries, material palettes, structural systems, construction details, etc.—with a degree of autonomy that goes beyond the straightforward designer-system input procedure and showing a commendable grade of autonomy during the generation process. Therefore, Generative AI systems, when tested with a prompt with a professional level of detail (e.g., a program description, a site boundary, a structural brief, a stylistic reference), generate families of candidate results that the professional can evaluate, test, or discard according to the coherence shown with the original prompted information.

This modality is one of the most significant advances towards the concept of autonomy in computer science and a detachment from the historical perception of the computer intended only as a mere executer and as a passive tool. Furthermore, this paradigmatic shift also changes the relationship between the designer and the AI model where the first is empowered in transforming himself from a simple form-maker into a proper creative director within a generative process [40]. From a productivity perspective the implication is groundbreaking: using an AI system, a designer can explore a larger design space in comparison with traditional methods, and can generate configurations and solutions that would have been harder to materialize without an AI counterpart.

Nevertheless, some limitations still need to be addressed in this modality. First, the generated outputs often suffer from violated constraints, spatial incoherence, and undesirable artefacts that need analogue identification and correction from the designer. Second, the direct translation from AI-imagery and digital fabrication procedures still remains challenging from a technical point of view (the so-called 'AI-to-manufacture-gap'). Third, the design authorship in these processes opens discussion concerning legal, ethical, and professional issues that are still not properly addressed from a legal perspective.

1.5 Integration with CAD/CAM Ecosystems and Additive Manufacturing Pipelines

To assess the utility of AI models for design and digital prototyping purposes it is fundamental not to understand their generative capacities into an isolated environment, but rather to analyse the extent to which their usage can be integrated with the most famous software and tools. This section examines the current state of AI integration with CAD/CAM ecosystems and additive manufacturing pipelines, and the principal technical challenges that still might limit this integration.

1.5.1 AI Integration Within CAD Platforms

Major commercial CAD platforms have begun integrating AI capabilities at multiple levels of the design workflow. Generative design features—currently the most commercially mature AI capability in CAD—enable users to set design constraints and objectives and receive AI-generated topology-optimized geometries within a CAD/CAM environment. Just to mention some of them, it is possible to find a generative design module in Autodesk™ Fusion™ 360: a topology optimization one in Siemens™ NX's or another generative design environment in PTC Creo™. Such systems try to bridge the gap between AI systems and CAD software through the creation of AI-generated geometries as solid models that can be directly annotated, edited, and then exported to manufacturing and prototyping systems.

Concerning API level and scripting, the integration of large language model capacities into CAD environment and API is one of the most interesting and promising fields of interest. LLMs are used to interpret natural language instructions that can be transferred to CAD to perform a series of operations—geometric features, constraint definition, building qualities, etc.—and can be found now in some of the most used commercial products [32]. It is the case of the Rhino.compute framework, the SolidWorks™ API, or the Autodesk™ Platform Services API that are developed with the aim of converting AI-based design intents into concrete operations and geometries [41]. One of the main gaps that still exists today in this research field is the representational distance between the CAD-native geometry and the AI-generated ones. Indeed, if the first one uses boundary representation (B-rep) solid geometry to perform precise engineering modelling, the AI ones rely on different kinds of outputs—such as point clouds or triangle meshes—that are not directly compatible with the other family of software. A manual conversion would need a complex process involving mesh repair, surface fitting, and topological reconstruction with the risk of approximation errors. Some of the strategies under development to overcome this issue foresees the development of CAD-native generative models that directly produce parametric construction sequences [32] and neural surface fitting for smoother and editable surfaces starting from AI point clouds [42].

1.5.2 AI in CAM and Process Planning

The integration of AI with Computer Aided Manufacturing (CAM) passes through the challenge of transitioning from validated design geometry towards optimized manufacturing instructions. In this field, AI application is moving around the automatic recognition of automated manufacturing operations from three-dimensional CAD models and the AI-assisted generation of CNC toolpaths [43].

For example, FBMs—Feature-based machining—are systems that are implementing machine learning to identify geometric elements—such as holes, pockets, bosses, and fillets—to associate them to specific machining strategies and have shown a good reliability and performance. Reinforcement learning approaches are used to optimize toolpaths through the training of specific agents that can discover efficient tool trajectories especially in cases such as complex sculptured surface where it is intrinsically difficult to apply analytical optimization [44].

All the researches are pointing towards the same direction: the technical idea of developing a seamless design-to-fabrication continuum where a specific set of information navigates continuously and interruptedly from conceptual generation to manufacturing execution with the lowest possible amount of manual/human intervention.

1.5.3 AI and Additive Manufacturing

Additive manufacturing (AM)—including Fused Deposition Modelling (FDM), Selective Laser Sintering (SLS), Stereolithography (SLA), and metal powder bed fusion—has shown synergic features in relation to AI-driven design. For instance, AM allows the fabrication of complex geometries while removing some of the manufacturability constraints (draft angles, undercuts, minimum wall thicknesses) that limit the use of AI in conventional subtractive manufacturing [45].

The application of AI in addictive manufacturing is happening at multiple levels such as the generation of specific geometries by topology optimization that respect AM's specific constraints like self-support requirements [46]: the correct prediction of AM process outcomes through trained regression models [47].

AI has been applied to additive manufacturing at multiple levels: the generation of AM-optimized geometries by topology optimization systems that respect AM-specific constraints such as self-support requirements [46]; the prediction of AM process outcomes (dimensional accuracy, surface roughness, porosity) from process parameters using trained regression models [47]; the automated generation of support structures to maintain structural coherence while minimizing support volume; the real-time monitoring in computer vision systems-trained printing defects [48].

The design-to-fabrication continuum—mentioned at the end of the previous subparagraph—is finding its promising realization precisely in the integration of AI design generation with AM fabrication. For example, a generative AI produces a complex and optimized geometry while a CAM system computes the optimized AM build orientation and support structure. After that, the AM performs the building procedure while the AI monitors to ensure adequacy with the original specification. This is just an example of an autonomous design-to-artefact pipeline driven by both AI tools and AM ones. At the current stage, even if fully integrated systems are still a matter of research for academics and professionals, their individual components show an adequate maturity that deserves attention and investment.

1.6 A Taxonomy of AI Models by Design Task Type

Starting from all the discourses conducted, this chapter aims to propose a structured taxonomy of the main AI model families based on the primary design task that each of them can perform. The idea is to offer to researchers and practitioners a practical tool to orientate and identify the correct AI tool for a specific design problem, rather than a comprehensive classification of the possible interactions a system can perform in the broad field of design.

This taxonomy revolves around six main design operations: (1) Form Generation—the creation of novel geometric configurations; (2) Performance Prediction—the relation between geometric outputs and representations and design performance; (3) Layout Optimization—how to arrange spatial elements to meet programmatic and performance-based requirements; (4) Style Transfer—the generation of specific aesthetic features or their transfer from an object to another; (5) Construction Sequence Generation—from design representation to fabrication-ready items; (6) Design Analysis and Classification—the understanding and categorization of design outputs.

In *Form Generation*, diffusions models and GANs are considered the most performative architectures, where diffusion models present more output fidelity, and GANs allow faster inference and better design manipulation in terms of latent space. For what concerns VAEs, they are promising in morphological exploration using latent space interpolation. In *Performance Prediction*, graph neural networks perform satisfying prediction tasks when the geometry is represented as a relational structure, while CNN-based surrogate models present rapid approximation of expensive simulation results (Finite Element Analysis-FEA, Computational Fluid Dynamics-CFD, daylighting).

In *Layout Optimization*, graph-conditioned GANs and reinforcement learning agents are effective from simpler floor plan generation to more complex combinatory spatial arrangements problems.

In *Style Transfer and Synthesis*, StyleGAN [49] and diffusion models have shown the most promising fine-grain results both regarding text condition and reference image guidance. In *Construction Sequence Generation*, an interesting direction is the one based on autoregressive transformer models that use CAD construction histories [32] since their outputs can be transferred into an editable format. In *Design Analysis and Classification*, the standard for image-based analysis procedures are still CNNs, even though transformer-based vision models (ViT [50]) are becoming increasingly competitive.

1.7 Benefits, Challenges, and Open Research Questions

1.7.1 Demonstrated Benefits

The recent years have collected many empirical evidence of the positive impact that the integration of AI-based technologies can bring into the fields of design and digital prototyping. Concerning productivity, many studies—spanning from architectural design, to design and engineering—showed how AI-assisted workflows can help humans to evaluate a large number of design options in a short amount of time and to perform high-fidelity simulations [36, 37]. In terms of design quality, the material efficiency can be improved from 30 to 70% while maintaining the same performance under specific stress or loading [35–46, 51–53]. Furthermore, AI models have confirmed their capability of suggesting non-intuitive design configurations—both in regards to spatial arrangements or geometry that rarely would be proposed through conventional design methods—that can respect performance requirements [40], and leading to design innovations in specific domains such as biomedical implant design or aerospace structural components since the AM removes some of the wrong manufacturing assumptions suggested by traditional design intuition.

1.7.2 Principal Challenges

Together with the mentioned benefits, there are still significant limits for the practical application of AI models in professional fabrication and design environments. First of all, data challenge: the level of fitness a proper training dataset requires is very difficult to collect in contexts where the design is proprietary and the design outputs are heterogeneous. Indeed, generating a series of design-relevant attributes (spatial quality, structural adequacy, aesthetic coherence) that are very likely situated in specific situations relies on subjectivity [54].

Furthermore, the interpretability challenge is also significant: for example, a topology-optimized geometry created by a surrogate model can surely be efficient, but the model cannot explain to the designers the specific element that guarantees its performance and that a structural engineer would definitely require to validate, modify, or extend the result [55]. On this topic, an active area of research is precisely the one that can enable AI models to produce meaningful and actionable design rationale together with their geometric outputs.

Nonetheless, the fabrication fidelity challenge—as mentioned before the existing gap between AI-based geometries and fabrication-ready models—still represents a constraint in contemporary design practice. Generative models are still not efficiently able to produce outcomes that can pass all the necessary engineering requirements. Indeed, the implementation of complex, multi-type constraint satisfaction is still an open research problem in AI-based prototyping pipelines [56].

1.8 Conclusion

This chapter aimed to offer to the reader a comprehensive review that could span from the most diffused AI model families, and their operative modalities, and the possible integration procedures that constitute the current field of research bridging together AI for design and digital prototyping. At the same time, it covered the historical path from rule-based expert systems to contemporary generative architectures—such as GANs, VAEs, transformers, and diffusion models—highlighting efficiencies and limitations for each of these families, and introduced a three-modality framework—optimization, augmentation, and generation—as a critical lens to understand the different trajectories through which AI models can intersect with design practice, and a structured taxonomy based on design task type as a practical tool for model selection.

Furthermore, the chapter focused on the CAD/CAM ecosystems and manufacturing pipelines and their possible relation with AI procedures, discussing with particular attention the current gap between AI-generated geometries and fabrication-ready solid models, which represent the real operative bottleneck for the realization of a continuous loop for design-to-fabrication pipelines powered by AI.

The following chapters of this volume are rooted on these foundations and investigate AI-driven design procedures using a progressively growing spatial scale: from product and object (Chap. 2), to the building and large-scale artefact (Chap. 3); from the urban territory (Chap. 4), to the platform and ecosystem landscape for AI-driven design and prototyping (Chap. 5) and the proposal of a methodological framework for a full design-to-fabrication continuum (Chap. 6). Moreover, this introducing chapter has to be intended as the conceptual vocabulary through which these scale-oriented investigations will be conducted, analysed, and compared to each other.

The central conviction animating this volume—and this chapter's contribution to it—is that AI's most consequential impact on design practice lies not in the autonomous replacement of designer intelligence, but in the amplification of human design capability: enabling the exploration of larger solution spaces, compressing evaluation and iteration cycles, and freeing designer attention for the contextual, ethical, and creative judgements that no current AI system is equipped to make. The realization of this potential requires not only continued technical development but a sustained commitment to human-centred integration—the design, not merely the deployment, of AI-augmented design workflows.

References

1. Sutherland IE (1963) Sketchpad: a man-machine graphical communication system. PhD thesis, Massachusetts Institute of Technology, Cambridge MA
2. Kyratsis P, Tzotzis A, Davim JP (2025) A review on CAD-based programming for design and manufacturing. CAD-based programming for design and manufacturing, 1–26
3. Tzotzis A, Manavis A, Efkolidis N, Kyratsis P (2021) CAD-based automated G-code generation for drilling operations. Int J Mod Manuf Technol 13:177–184. https://doi.org/10.54684/ijmmt.2021.13.3.177
4. Goodfellow I, Pouget-Abadie J, Mirza M et al (2014) Generative adversarial nets. In: Advances in neural information processing systems (NeurIPS), Montreal, pp 2672–2680
5. Nauata N, Hosseini S, Chang KH, Chu H, Cheng CY, Furukawa Y (2021) House-gan++: Generative adversarial layout refinement network towards intelligent computational agent for professional architects. In: Proceedings of the IEEE/CVF conference on computer vision and pattern recognition, pp 13632–13641
6. Vlassis N, Ma R, Sun W (2020) Geometric deep learning for computational mechanics part I: anisotropic hyperelasticity. Comput Methods Appl Mech Eng 371:113299. https://doi.org/10.1016/j.cma.2020.113299
7. Huang Y, Zheng H, Cong M et al (2017) CAD-based automatic path generation and optimization for laser cladding robot in additive manufacturing. Int J Adv Manuf Technol 92:3605–3614. https://doi.org/10.1007/s00170-017-0384-0
8. Liu Z, Wald I, Poole E et al (2023) OpenShape: scaling up 3D shape representation towards open-world understanding. In: Advances in neural information processing systems (NeurIPS), New Orleans
9. Reddy EJ, Reddy KS, Reddy PN (2024) Recent advancements in knowledge-based parametric modeling of mechanical components. In: Yadav S et al (eds) Proceedings of third international conference in mechanical and energy technology. Springer Nature, Singapore, pp 401–411
10. Gero JS (1990) Design prototypes: a knowledge representation schema for design. AI Mag 11(4):26–26
11. Holland JH (1992) Adaptation in natural and artificial systems: an introductory analysis with applications to biology, control, and artificial intelligence. MIT Press
12. Hinton GE, Osindero S, Teh Y-W (2006) A fast learning algorithm for deep belief nets. Neural Comput 18(7):1527–1554. https://doi.org/10.1162/neco.2006.18.7.1527
13. Krizhevsky A, Sutskever I, Hinton GE (2012) ImageNet classification with deep convolutional neural networks. In: Advances in neural information processing systems (NeurIPS), Lake Tahoe, pp 1097–1105
14. Kingma DP, Welling M (2014) Auto-encoding variational Bayes. In: Proceedings of the international conference on learning representations (ICLR), Banff
15. Vaswani A, Shazeer N, Parmar N et al (2017) Attention is all you need. Advances in neural information processing systems, 30
16. Brown T, Mann B, Ryder N et al (2020) Language models are few-shot learners. In: Advances in neural information processing systems (NeurIPS), virtual, pp 1877–1901
17. Lee JDMCK, Toutanova K (2018). Pre-training of deep bidirectional transformers for language understanding 3(8):4171–4186. arXiv preprint arXiv:1810.04805
18. Radford A, Kim JW, Hallacy C et al (2021) Learning transferable visual models from natural language supervision. In: Proceedings of the international conference on machine learning (ICML), virtual, pp 8748–8763
19. Ramesh A, Dhariwal P, Nichol A et al (2022) Hierarchical text-conditional image generation with CLIP latents. arXiv:2204.06125
20. Rombach R, Blattmann A, Lorenz D et al (2022) High-resolution image synthesis with latent diffusion models. In: Proceedings of the IEEE/CVF conference on computer vision and pattern recognition (CVPR), New Orleans, pp 10684–10695. https://doi.org/10.1109/CVPR52688.2022.01042

21. Ho J, Jain A, Abbeel P (2020) Denoising diffusion probabilistic models. In: Advances in neural information processing systems (NeurIPS), virtual, pp 6840–6851
22. Chaillou S (2022) Artificial intelligence and architecture: from research to practice. Birkhäuser
23. Zhou B, Lapedriza A, Khosla A et al (2018) Places: a 10 million image database for scene recognition. IEEE Trans Pattern Anal Mach Intell 40(6):1452–1464. https://doi.org/10.1109/TPAMI.2017.2723009
24. Ronneberger O, Fischer P, Brox T (2015) U-Net: convolutional networks for biomedical image segmentation. In: Navab N et al (eds) Medical image computing and computer-assisted intervention (MICCAI). Springer, Cham, pp 234–241. https://doi.org/10.1007/978-3-319-24574-4_28
25. Mirza M, Osindero S (2014) Conditional generative adversarial nets. arXiv:1411.1784
26. Isola P, Zhu J-Y, Zhou T, Efros AA (2017) Image-to-image translation with conditional adversarial networks. In: Proceedings of the IEEE conference on computer vision and pattern recognition (CVPR), Honolulu, pp 1125–1134. https://doi.org/10.1109/CVPR.2017.632
27. Zhu J-Y, Park T, Isola P, Efros AA (2017) Unpaired image-to-image translation using cycle-consistent adversarial networks. In: Proceedings of the IEEE international conference on computer vision (ICCV), Venice, pp 2223–2232. https://doi.org/10.1109/ICCV.2017.244
28. Newton D (2019) Generative deep learning in architectural design. Technology| Architecture+ Design 3(2):176–189
29. Chang AX, Funkhouser T, Guibas L et al (2015) ShapeNet: an information-rich 3D model repository. arXiv:1512.03012
30. Nauata N, Chang K-H, Cheng C-Y et al (2020) House-GAN: relational generative adversarial networks for graph-constrained house layout generation. In: Vedaldi A et al (eds) Computer Vision—ECCV 2020. Springer, Cham, pp 162–177. https://doi.org/10.1007/978-3-030-58452-8_10
31. Zhao H, Jiang L, Jia J et al (2021) Point transformer. In: Proceedings of the IEEE/CVF international conference on computer vision (ICCV), Montreal, pp 16259–16268. https://doi.org/10.1109/ICCV48922.2021.01595
32. Willis KDD, Jayaraman PK, Lambourne JG et al (2021) Engineering sketch generation for computer-aided design. In: Proceedings of the IEEE/CVF conference on computer vision and pattern recognition workshops (CVPRW), virtual, pp 2105–2114
33. del Campo M (Ed) (2024) Artificial intelligence in architecture. John Wiley & Sons
34. Poole B, Jain A, Barron JT, Mildenhall B (2022) DreamFusion: text-to-3D using 2D diffusion. arXiv:2209.14988
35. Sigmund O (2001) A 99 line topology optimization code written in Matlab. Struct Multidiscip Optim 21(2):120–127. https://doi.org/10.1007/s001580050176
36. Regenwetter L, Curry B, Ahmed F (2022) Deep generative models in engineering design: a review. J Mech Des 144(7):071703. https://doi.org/10.1115/1.4053859
37. Li J, Du Q, Feldman J (2022) Surrogate-based aerodynamic shape optimization with the active subspace method. Struct Multidiscip Optim 59(2):403–419. https://doi.org/10.1007/s00158-018-2073-5
38. Gao R, Sanghi A, Shugrina M et al (2022) CLIP-Forge: towards zero-shot text-to-shape generation. In: Proceedings of the IEEE/CVF conference on computer vision and pattern recognition (CVPR), New Orleans, pp 18603–18612. https://doi.org/10.1109/CVPR52688.2022.01805
39. Kakoulis KG, Konstantinidis P, Manavis A, Kyratsis P (2023) A system for automatic dimensioning of mechanical drawings. Int J Mod Manuf Technol 15(3):67–74. https://doi.org/10.54684/ijmmt.2023.15.3.67
40. Oxman R (2017) Thinking difference: theories and models of parametric design thinking. Des Stud 52:4–39. https://doi.org/10.1016/j.destud.2017.06.001
41. Tzotzis A, Manavis A, Efkolidis N, Kyratsis P (2024) Modeling the strength-to-mass ratio of the fused filament fabricated abs polymer with ANN. Int J Modern Manuf Technol 16(2):7–12
42. Maturana D, Scherer S (2015) VoxNet: A 3D convolutional neural network for real-time object recognition. In: Proceedings of the IEEE/RSJ international conference on intelligent robots and systems (IROS), Hamburg, pp 922–928. https://doi.org/10.1109/IROS.2015.7353481

43. Zhang X, Nassehi A, Newman ST (2014) Feature recognition from CNC part programs for milling operations. Int J Adv Manuf Technol 70:397–412. https://doi.org/10.1007/s00170-013-5275-4
44. Peddireddy D, Fu X, Wang H et al (2020) Deep learning based approach for identifying conventional machining processes from CAD data. Procedia Manuf 48:915–925. https://doi.org/10.1016/j.promfg.2020.05.130
45. Koul P (2024) A review of generative design using machine learning for additive manufacturing. Advances in Mechanical and Materials Engineering, 41(1):45–159. https://doi.org/10.7862/rm.2024.14
46. Sigmund O, Maute K (2013) Topology optimization approaches: a comparative review. Struct Multidiscip Optim 48(6):1031–1055. https://doi.org/10.1007/s00158-013-0978-6
47. Huang J, Kwon O-Y, Oh J et al (2023) Diffusebot: Breeding soft robots with physics-augmented generative diffusion models. Adv Neural Inf Process Syst 36:44398–44423
48. Patel S, Mekavibul J, Park J, Kolla A, French R, Kersey Z, Lewin GC (2019) Using machine learning to analyze image data from advanced manufacturing processes. In: 2019 systems and information engineering design symposium (SIEDS), pp 1–5. IEEE
49. Karras T, Laine S, Aila T (2019) A style-based generator architecture for generative adversarial networks. In: Proceedings of the IEEE/CVF conference on computer vision and pattern recognition (CVPR), Long Beach, pp 4401–4410. https://doi.org/10.1109/CVPR.2019.00453
50. Dosovitskiy A, Beyer L, Kolesnikov A et al (2021) An image is worth 16x16 words: transformers for image recognition at scale. arXiv preprint arXiv:2010.11929.
51. Kyratsis P, Tzotzis A, Markopoulos A, Tapoglou N (2021) Cad-based 3d-fe modelling of aisi-d3 turning with ceramic tooling. Machines 9(1):4
52. Attene M, Livesu M, Lefebvre S et al (2015) Shapes in a box: Disassembling 3D objects for efficient packing and fabrication. Comput Graph Forum 34(8):64–76. https://doi.org/10.1111/cgf.12608
53. Fountas NA, Vaxevanidis NM, Stergiou CI, Benhadj-Djilali R (2019) Development of a software-aided environment for optimizing CNC sculptured surface machining. Int J Adv Manuf Technol 104:1–20
54. Bommasani R, Hudson DA, Aditi E et al (2021) On the opportunities and risks of foundation models. arXiv:2108.07258
55. Lipton ZC (2018) The mythos of model interpretability. Queue 16(3):31–57. https://doi.org/10.1145/3236386.3241340
56. Constante-Flores GE, Chen H, Li C (2025) Enforcing hard linear constraints in deep learning models with decision rules. arXiv preprint arXiv:2505.13858.

Chapter 2
The SMALL(S): AI for Product and Object Design

Abstract The first chapter introduced the theoretical foundation and structure taxonomy at the core of this AI-driven exploration intersecting the field of design and digital prototyping while, at the same, proposing a critical mapping of the principal families of AI models subdivided into six main design task types. The current chapter builds on that general framework and starts exploring the different spatial scales presented in this volume starting from the tiniest and most intimate one: the ones dealing with product and object design. The objects presented here are intended from a typological perspective—chairs, shoes, luminaires, tableware, wearables, and bicycles—and from the domain of their everyday usage addressing the domestic, working, and public life dimension. The chapter aims to intersect the peculiar design challenges at this scale—the connection between aesthetic form and structural performance; the balance between standardized and customized solutions, etc.—and discusses the methodological and instrumental development that is shaping the use of AI models in this field. The examples depicted are taken from the existing literature in the fields of furniture, footwear, and consumer product customization with the aim to ground the discussion in concrete applications. The conclusion serves to highlight a series of benefits, limitations, and future research trajectories related with the impact of AI in their design processes.

Keywords Generative design · Product design · Object design · Furniture design · Footwear design · Topology optimization · Diffusion models · Variational Autoencoders · Mass customization · Additive manufacturing · Design automation

2.1 Introduction

The taxonomy proposed in Chap. 1 aimed to systematize the burgeoning field of AI-driven design through the use of six main task types while, at the same time, highlighting the most appropriate AI model families for each one of them. Moreover, it discussed how design practices intersect a broad hierarchy of different spatial scales—from individual objects and components to large-scale construction artefacts

P. Kyratsis et al., *AI Models for Design and Digital Prototyping*, Manufacturing and Surface Engineering, https://doi.org/10.1007/978-3-032-35144-9_2

and the urban scale—and the methodological and technical challenges derived from AI implementation at each of those. The current chapter investigates the tiniest and most intimate one of these scales: the designed object one.

The designed object can be considered the oldest and most fundamental entity of design practices. It connects together all those artefacts—such as chairs, shoes, lamps, and bicycles—where structural and materiality features, aesthetic intention, and types of uses are in relation with the human body and the human hand. At this scale, design is shaped by a tension which is very difficult to identify at larger scales and that always oscillates between the measurability of a performance (a chair must support its load and a bicycle frame must resist fatigue and stress) and the inner relation between form and function. Indeed, an aesthetically pleasing chair is not necessarily a solid chair and vice versa.

Human designers add to the design process a systematic and embodied aesthetic intelligence accumulated, thanks to centuries of making, looking, and using, and dimension of design judgement that cannot be addressed by a sole engineering formula and that relates with surface quality, proportional system, form language, and material expression.

AI models have tried to engage with both these dimensions of object design: the aesthetic and the performative one. On the form side, generative adversarial networks, variational autoencoders, and diffusion models have shown a promising capacity to explore, generate, and interpolate object forms across different scales and with a diversity that not a single designer mind could bear alone. On the performative side, topology optimization systems and neural surrogate models displayed the capacity of imagining structural configurations that also an experienced engineer would find challenging to achieve through conventional tools. If Chap. 1 discusses the optimization and generation modalities of AI design procedures, the current one delves deeper into the matter and demonstrates how they operate at the object scale and how in contemporary workflows designers have started to combine them together.

This chapter is organized around the following structure: Sect. 2.2 discusses the object scale as a proper design domain and the specific properties for which AI-driven strategies can be applied. Section 2.3 focuses on structural and mechanical performance optimization in the age of AI. Section 2.4 expands the possibilities of aesthetic explorations and form generation through AI. Section 2.5 delves into the integration of aesthetic and structural approaches for compound object design. Section 2.6 discusses the newer possibility in terms of mass customization and personalization. Section 2.7 analyses the challenges for AI-generated forms to be fabrication-ready geometries. Section 2.8 focuses on benefits, limitations, and open research directions. Finally, Sect. 2.9 represents a conclusive bridge to Chap. 3.

2.2 The Object Scale: Characteristics and Design Challenges

In order to examine the structure and entity of design problems at this scale, it is fundamental to understand what makes object design unique—and how this unicity affects AI workflows and tools. Such problems gather inside of them a series of properties that all together might determine the feasibility of AI integration and also the risks and critical issues this might cause.

2.2.1 The Inseparability of Structure and Aesthetic Form

In regards to the object scale, it is impossible to separate aesthetic form and structure performance and to think that these categories might be optimized independently and then combined. For example, the cross-section of the leg of a chair is at the time an aesthetic feature that gives visual character to the object and a structural device that have to resist to stress and bending, while the shoe sole is a surface geometry that only gives an aesthetic profile to the product but it also related to cushioning performance. Considering this, it is clear how AI models only dedicated to structural optimization—topology optimization systems, neural surrogate models—would take the risk to produce structurally efficient but aesthetically incoherent results [1].

This close relationship is indeed a primary challenge for AI-assisted object design while not appearing with the same intensity at a larger scale. For example, if an aerospace bracket which is perfectly structurally optimized does not need to be beautiful, a structurally optimized chair definitely has also not only to be aesthetically pleasant but also comfortable, culturally meaningful, and the expression of a design intention. Indeed, one of the central methodological challenges at the object scale is the development of AI workflows that can engage both the structural and aesthetics dimension of design practices.

2.2.2 Material as Form: The Object's Expressive Dimension

Materiality is not to be considered just a mere performance parameter at this scale. It represents an ontological dimension of the object intersecting its own aesthetic and cultural identity. The tactile sensation given from a shoe sole, the translucently of polycarbonate in a light design product, or the wood roughness in a furniture piece are qualities that do not just follow mechanical property topology optimization systems, and for this reason they tend to be very difficult to be optimized by any existing AI. Such characteristics are the result of centuries of material science, craft knowledge, and aesthetic judgement that the human designer brings with him consciously and unconsciously into any design operation he performs [2]. Even though AI models have improved in creating meaningful associations between material qualities and aesthetic features—diffusion models have been trained on dataset to generate outcome to coherently render different materials on the same shape—they still have to show if they can properly reason about materials as a design medium as a professional would do. This limitation is particularly decisive in generation modality: AI objects usually lack what is called material intelligence—the idea that a form should be shaped by and for the material from which it is made—and that serves to identify compelling designed objects from just geometrically interesting shapes.

2.2.3 The Domestic Scale and the Human Body

The majority of objects are imagined and designed for a specific use by the human body in domestic and working environments. In this regard, an important set of constraints concerning object design are the dimensional and ergonomic requirements that are always very difficult to correctly formalize but also very easy to be violated. The challenging of this formalization relates to needed scale of precision—millimeters—and the domain complexity—three-dimensional human body variation across diverse populations—at which these ergonomic constraints operate. For example, 5 mm difference in a chair seat can create user discomfort, while a shoe that does not accommodate differences concerning the width of a foot population is automatically translated into a commercial failure despite its aesthetic or structural qualities [3]. To overcome such issues, specific AI models trained on large datasets of human body measurements and proportions—e.g., statistical shape models, deep learning regression networks—have shown interesting abilities to impact ergonomic constraints in object design [4]. If proficiently integrated in design pipelines, they open up to the possibility of creating geometries that can ergonomically adapt to different populations without losing their structural and aesthetic coherence. The counterpart of this ability—called individualized personalization: the capacity of design object for specific individual rather than for an average population—will be presented in Sect. 2.6 as one of the most interesting near-term opportunities for AI in object design.

2.3 AI-Driven Structural Performance Optimization of Designed Objects

In object design, the structural performance dimension is the requirement for the objects to bear loads, to resist stress and fatigue, and to maintain their geometric integrity for their whole design life. Structural performance is currently the category giving the best results in terms of AI-driven optimization. This section will review some of the main tools that conceptually refer to the optimization modality introduced in Chap. 1, with the difference that even though their application in object design follows rules that are very similar to engineering components application, in this case aesthetic features must be considered alongside structural performance.

2.3.1 Topology Optimization for Object Structures

Topology optimization—the optimal material distribution to guarantee optimal structural performance to objects affected by loading and boundary conditions—has become a primary field of research in object design (from design furniture to consumer products) since structural efficiency has a direct impact on material cost, product mass, and user perceived quality [5]. Branching, lattice-reinforced, or membrane structures that concentrate material along main stress path are just some of the most common outputs of topology optimization and have demonstrated aesthetically productive in contemporary furniture and product design since their peculiar organic forms have been interpreted as a distinct and successful visual language.

The work of the Dutch design Jori Laarman and his series of bone-growth and gradient furniture structures—generated through form-finding algorithms and reified in materials such as steel, marble, and resin—testifies an early attempt to turn topology optimization from a mere engineering tool to a proper design methodology [6]. In his research, Laarman used bone-growth principles to train generative algorithms that he then applied to the form of domestic chairs and tables, with the results of producing not only optimized structural topologies but also object with a meaningful aesthetic proposition. Laarman artefacts represent a formal vocabulary—based on an inner computational logic—that has no precedent in industrial design. For the first time, topology optimization is not a hidden quality of an object but an expressive mean to define its primary formal identity. Generative tools such as Autodesk™ Fusion™ 360, nTopology™, and Rhinoceros™ with Grasshopper™ plug-ins have incentivized the practical deployment of topology optimization in object design and made it accessible also to designers without any specific computational expertise [7]. These tools allow professionals to specify domains, load cases, and manufacturing process constraints, while at the same time producing structurally optimized geometries that can be assessed—and eventually modified—in a familiar design workflow and environment. Moreover, their integration in manufacturing fabrication has been one of most groundbreaking integrations in the field of object design in the last 20 years.

2.3.2 Surrogate Models and Real-Time Structural Feedback

The application of neural network surrogate models—networks trained to approximate the outputs of computationally expensive structural simulations—to object design has allowed design to perform formal decisions while receiving real-time structural feedback during the design process. Nie et al. [8] have showed the potentiality of convolutional neural networks—if trained on finite element analysis—to produce results able to provide instant calculations concerning maximum deflection and stress distribution when a design operates form adjustments, generating a feedback loop between structural consequences and formal decision that now needs seconds and not hours to be performed.

Such real-time feedback represents an important novel for object design because it enables the designer to manage structural and aesthetic judgement almost simultaneously and not anymore as a linear sequence. Indeed, instead of separating formal and structural decision-making, the designer can explore this relationship simultaneously and explore form space always with the necessary structural awareness. This capability is not only significance for its productivity application but also for its educational one in the nurturing of new generation of designer able to understand form and structural performance in a holistic way of thinking.

2.4 AI-Driven Form Generation and Aesthetic Exploration

The creation and exploration of novel objects form—that have the aim of expanding the formal language of a well-established typology, blending different typologies together, or exploring new direction for design practice—are undoubtedly the dimension of AI-assisted object design that refers to the generation modality described in Chap. 1. All the AI model families presented in this part—from generative adversarial network, to variation autoencoders, to diffusion modes—have their own peculiar approach to addressing this challenge, and different implications concerning the way the designer is engaged in the process.

2.4.1 Learning Object Typologies: GANs and the Design Corpus

When trained on a specific database, generative adversarial networks (GANs) learn how to statistically represent object typologies. Indeed, they are able to understand the implicit rules behind the generation of a form, and the characteristics and detail that define what a chair, table, shoe, etc. is, and to generate novel artefacts within the typologies they identify that can also differ from any given specific examples from the training set. A generative adversarial network, which is trained on a database of three-dimensional chair models, can generate design options that are clearly referable to a chair and, at the same, showcase some formal gestures, proportional relationship, or structural configuration that no one of the studied chairs had [9].

GANs contribution does not fall specifically in the field of autonomous generation of finished and ready-to-be-printed designs—all the outputs require a certain amount of human intervention—but their implementation is definitely a game-changer in the early-stage ideation and conceptualization phase. Furthermore, the role of the designers is transformed from being the primary form-maker to an artistic director and conductor of a generative process, as it was also mentioned in Chap. 1 when the idea of the reconfigured role was presented as a consequence of generation-modality in design practices empowered by AI.

The subfamilies of Conditional GANs (cGANs), where the generation can receive an auxiliary conditioning in terms of desired attributes required to the output, further expand the generative possibility and focus them not on the all-product shape but specific aesthetic regions [10]. For example, a cGAN with a robust training on style attributes—minimalist versus baroque, rectilinear versus curvilinear, etc.—is able to generate object forms that position themselves in a specific region of the aesthetic space and research, and allow designers to investigate specific formal territories.

2.4.2 Navigating Design Space: Variational Autoencoders and Morphological Exploration

Variational Autoencoders (VAEs) propose a complementary approach that allows form generation processes to perform a systematic navigation of a design space they have learnt through training. When a VAE is trained with a family of design objects it comprehends that every point in the latent space is related to an object form and a continuous path between these points corresponds to design configurations that undergo a smooth morphological transformation [11].

Two are the design workflows that are triggered from this latent space navigation and play a role of interest in object design. Interpolating two existing designs corresponds to a morphological sequence that smoothly generates transitions between them, and facilitates for the designer the exploration of all the formal nuances lying in the territory between the original references. Attribute manipulation, which relies on identifying the latent direction that is more associable to a specific formal attribute and exploring the space in that direction, empowers designers to focus on specific features of the object form while keeping all the others ones constant and untouched.

Umetani [11] report a representative application of VAE explorations in the field of automotive design. The researcher trains a VAE on a dataset of three-dimensional car models from the ShapeNet™ database [9] and used the trained model as an interface for design exploration. When the designers navigate the learnt latent space from the model—in the transition zones between well-populated typological cluster—they encounter formal configurations with attributes drawn from mid-century and contemporary design that were not embodied by any of the existing object in the dataset. These configurations are described by the authors as formally productive since they remain outside of the repertoire that conventional design training aims to make readily accessible. This very ability to identify meaningful and formally coherent configurations outside of the domain of individual design memory is one of the most interesting benefits of VAE-based exploration in product design, and connects to the observation that AI-generated creations can expand the designer agency through a qualitatively different mechanics from conventional ideation processes.

2.4.3 *Diffusion Models for Object Concept Visualization*

Text-condition image diffusion models—such as Stable Diffusion™, DALL·E™ 2, and Midjourney™—are mostly implemented in product and object design practice as instruments for early-stage idea and concept visualization. Their ability to generate photorealistic renders from natural language-based prompts has enhanced designers in quickly exploring design options and communicate formal directions since the concept development stage without embarking in time-expensive processes of three-dimensional modelling. Concerning limitations, many of the images that this AI-family produces present a large number of geometric inconsistencies—individual views that cannot be intended as a coherent three-dimensional form, structural configurations that are aesthetically pleasing but physically impossible, and that do not follow any of the existing rules concerning structural performance, etc.—that penalizes designers that attempt to translate this imagery into a three-dimensional model [12]. Such inconsistencies are part of the fundamental nature of image diffusion models which is predominantly effective exclusively in terms of two-dimensional outputs and related to the identified representation gap.

Nevertheless, there are some specific three-dimensional shape diffusion models that have the objective to address this limitation through the direct generation three-dimensional geometries. Among the preferred designers' choices there is LION™ [13]—a latent point diffusion model developed by Vahdat et al.—and Shap-E™ [14] by Jun and Nichol. Both of them are capable of generating three-dimensional object forms from natural language text or image conditioning and to generate useful outputs for early-stage form exploration, even though still not at the correct resolution for a direct manufacturing integration. However, the development curve for three-dimensional diffusion models is quite rapid, and their role in early-stage object design ideation and conceptualization is already well-established within the design community.

2.5 Integrating Structural and Aesthetic AI: Compound Object Design

As discussed at the beginning of this chapter, the eventual separation during the design process between structural optimization procedures and aesthetic form generation has the risk to not respect one of the fundamental characteristics of design practice:

the one that says that in the most compelling design object, there is a total coherence between aesthetic form and structure. Throughout the history of design, the best furniture are not remembered because they present a structure that supports an aesthetic: but because the aesthetic itself is the structure with this inner relationship that is made by the designer visible and expressive.

For example, the best shoes do not showcase a performance sole and aesthetic upper: the performative qualities of the sole—cushioning gradient, its torsional flexibility, surface articulation, etc.—are perfectly assimilated with the shoes' visual identify. This section delves into an AI workflow that addresses aesthetic and structure as an integrated design problem.

2.5.1 Sequential Multi-Modality Workflows

Currently, the integration of AI-based structural and aesthetic features in object design mostly follows the path of a sequential multi-modality workflows, in which the outputs of one phase are used as input or constraint for the following ones. A canonical sequence starts from the concept generation through diffusion models—to produce visual outcomes that could set the ground for the aesthetic domain of the process—that could explore a large spectrum of form configurations coherent with the selected aesthetic direction and it is then concluded by a final topology optimization to meet the performance requirements and constraints prior to the manufacturing phase. It is clear how each one of these transitions requires a human judgement: the human intelligence evaluates, selects, and steers the process between the two modalities, and provides some specific contextual inputs that no AI model embeds [15].

The primary source of inefficiency recorded in sequential workflows is the representation gap between modalities, that is the mismatch between the output representation produced by one of the AI models and the specific input by the next. In fact, diffusion models produce two-dimensional images; VAE inputs are outputs that belong to the family of three-dimensional point clouds; and topology optimization requires solid Computer Aided Design (CAD) geometry to be performed. Each one of the transitions between these formats requires a conversion process that demands manual intervention and that it is not safe from approximation errors. For this reason, an active research frontier is the development of unified representations that can be generated, navigated, and optimized among the different modalities to reduce sequential workflow friction and inconsistencies [16].

Salles and Gyi [3] present a well-documented example of this sequential approach in regard to footwear design and then elaborated in further studies focusing AI-assisted sole development. The overall workflow—diffusion-conditioned concept images generation, three-dimensional form navigation using a VAE trained on sole geometry datasets, and a final multi-load-case topology optimization within fabrication constraints specified for selective laser sintering of thermoplastic polyurethane [9]—demonstrated the possibility to produce sole geometries where energy return and mass performance are way higher that the ones achievable by conventional design process [5]. As will be discussed in Sect. 2.7, the reported bottleneck is the last transition from VAE to topology optimization where point cloud outcomes require significant manual reconstruction time where possible errors and mismatches might emerge.

2.5.2 *Performance-Informed Generative Design*

The integration of performance constraints it is fundamental since the generative model's training represent an alternative to sequential multi-modality workflows in order to produce AI models that can generate aesthetically coherent forms that can also satisfy the requirements in terms of structural performance. Vlassis et al. [17] discuss a geometric deep learning approach where a neural network has been trained on a finite element simulation dataset to learn to independently generate structural geometries that can fulfil multiple performance criteria with the generation process that does not need a subsequent post-generation optimization phase. Physics-informed neural networks (PINNs), introduced by Raissi et al. [18], follow this direction and are empowered by structural mechanics equations since the neural network training, enabling surrogate generation of geometries that are adherent physical laws even for configurations not represented in the training data.

From a technical perspective, it is important to underline that performance-informed generative design is more demanding that sequential multi-modality workflows because it has to compute a huge amount of performance metrics in relation to the generated geometry. Current researches have tested this approach for parametric frames, shell structures, and lattice components, while more complex geometries and topologies still remain a limitation.

Nevertheless, the possibility to collapse sequential workflows into a single generative step—able to produce aesthetically coherent and structurally adequate outputs—makes this direction one of the most significant and promising open directions for AI-based research in object design.

2.6 AI-Enabled Object Customization and Mass Personalization

Mass personalization is surely one of the most commercially successful application of AI generative process at the object scale since it allows to intersect individual customer specifications—functional, aesthetic, dimensional, etc.—at a speed closer to the one concerning standardized mass production. In this regard, the application of AI families in this field addresses a long-time lasting tension existing in object design between the experiential superiority of individual tailor-made object and the economic logic that lies behind standardization procedures.

2.6.1 Parametric Design Systems and AI Interfaces

From a technological perspective, parametric design system is the technology at the core of AI-assisted object customization: it can be described as a computation model where the geometries are defined through a function of a set of adjustable and modifiable parameters where any change to a valid parameter generates a chain reaction of geometrically coherent, structurally sound, and manufacturable object variant. Considering this, it is clear how CAD-based programming approaches—built using the APIs of widespread commercial software such as SolidWorks™, Autodesk Inventor™, and Rhinoceros™ with its visual programming language plug-in Grasshopper™, represent the base from which implement parametric object models spanning from furniture to footwear, from lightning to accessory categories [19, 20].

In this specific field, AI models can serve as a proper intelligent layer of interface between customer input and parametric design systems: they can translate and interpret natural language descriptions, visual references, dimensional measurements, and apply a preference selection to build the parameter specifications that will conduct the geometric outcomes. If a customer described a dining chair as 'low and wide, with a wood seat and a light, airy base', he will be providing a series of information that the AI language model, previously trained on a design vocabulary associated with a parametric environment's parameter space, can immediately convert into some parameter categories that the parametric system can use to produce a series of candidate designs. The subsequent customer reviews will provide further feedbacks that the AI interface can elaborate and put into some new parameters to restart the cycle until a final and satisfactory configuration is achieved.

The commercial potential of this direction is demonstrated by Manavis et al. [20], who focused on the elaboration, implementation, and deployment of a mass-customization system for residential furniture created combining together a Grasshopper-based parametric engine, a SolidWorks™ API-based manufacturing preparation module. The parametric model is divided into Code_1 and Code_2 that performs separate but intertwined operations.

Code_1 is divided into two main sections, A and B. While section A it used to generate the flat surfaces and shelves for the two different desks, section B is dedicated to the creation of the legs and connections between the shelves. The Code_2 algorithm is used to generate furniture assembly instructions based on the output data of the final application. At the end of the process, a fully complete building manual is generated and ready to be used from the customer. Again, in this process the user feedback is of extreme importance because it can be immediately converted into individual design options from the Grasshopper™ script at the core of the experiment that can then be converted both into a building manual or a.STL file ready to manufacturing. The implementation of AI families into such consolidated processes is one of the open research directions to pursue the balance between mass customization and mass personalization dynamics and bridge the representation gap.

2.6.2 *Anthropometric Customization: Objects Shaped by Bodies*

Besides providing new research directions in terms of aesthetic, structural, and dimensional customization, AI also allows a meaningful adaptation of object geometry to individual body measurements and characteristics, and the generation of artefacts that are totally fitted in relation to the human body—like, for example, footwear lasts adapted to the customer's foot geometry or bicycle frames perfectly sized to the rider's body position and proportions. In this regard, Statistical Shape Models (SSMs)—probabilistic models trained on large datasets of three-dimensional body scans to interpret shape variation in a population—represent a computation foundation for the so-called anthropometric customization [4]. Starting from a set of measurements from a three-dimensional scan of an individual bodily portion, an SSM is able to create a personalized geometric template that embeds that individual's morphological features. Salles and Gyi [3] have tested this approach to personalized insoles via addictive manufacturing, and demonstrate that SSM-based customization provides a better pressure distribution in comparison to standardized solution for a population of individuals differing in terms of foot morphologies. From an economic point of view, the use of this methodology has been facilitated by a better accessibility to three-dimensional scanning hardware, where smartphone-based scanning applications are now good enough to generate with sufficient accuracy set of data that be used for anthropometric customization.

2.6.3 *Generative Design for Product Family Coherence*

Coherent product families are a collection of related objects that address different functional requirements while maintaining a shared formal vocabulary. To make it clearer, a collection of furniture might include a lounge chair, a kitchen chair, and armchair, and a stool where all of them express the same design language but differ in function, proportions, or structural systems; a lightning collection can include floor, table, and wall variants that are recognizable and formally coherent despite their different functional objective.

Style-based generative architectures, like the StyleGAN framework proposed by Karras et al. [21], allow the transfer of specific formal identity between different object typologies through the separation of the representation of content and style in the generative model. Such models can generate new typologies that inherit the formal characteristic of the original collection without the need for manual specification for every design step. StyleGANs allow the design teams to maintain a formal coherence among objects that would be hardly achievable with traditional methods, especially when the family's formal logic is not articulated in design brief but rather implicit and deductible only from the already existing member of the family.

2.7 From AI-Generated Form to the Fabricated Object

The 'AI-to-manufacture gap'—that is the distance between the outcomes produced by AI generative models and the level of readiness in geometries required by manufacturing processes—can be recognized at the object scale in different modalities that reflect both the specific characteristic of object geometry and also the heterogeneity of processes through which these objects are produced and manufactured.

2.7.1 *Object Geometry and Fabrication Process Compatibility*

In the design field, objects are produced through a very wide range of fabrication and manufacturing processes—CNC routing of sheet materials, injection moulding, vacuum forming, casting, hand forming, additive manufacturing, craft finishing, etc.—that pose strong constraints on the geometries they are able to understand thus to produce. In terms of compatibility, there is a broad spectrum of differences when it comes to AI-generated designs: additive manufacturing is undoubtedly one of the most compatible technologies for its capacity to produce geometries of arbitrary complexity; injection moulding is the least compatible procedures—since it requires variable-density characteristic of topology optimization outputs [22].

For this reason, the selection of the correct fabrication process should be a constraint specified since the beginning of the generation process and not a downstream consideration to be taken after the finalization of AI-generated form. Current practices of topology optimization and generative design tools support the integration of manufacturing process constraints—such as additive self-support angle thresholds, mould draft angle requirements, CNC tool accessibility constraints—as a hard condition during the generation phase [7]. Sbrugnera Sotomayor et al. [22] developed a systematic review of CAD based on pre-processing tools that deals with this integration challenge for additive manufacturing, and showed how a series of hard-constraint during the generative loop sensibly reduces the need for manual correction before initiating the manufacturing phase.

2.7.2 *Representation Conversion at the Object Scale*

The representation gap—the mismatch between the output of AI generative models (point clouds, triangle meshes, neural implicit surfaces) and the boundary representation for solid geometries which is required by manufacturing technologies—can be found at the object scale in forms that testify the particular characteristics of object geometry. For example, object surfaces are typically smooth and show continuous curvature—a chair shell, a shoe last, a luminaire body—rather than faceted, and this situation usually makes the surface fitting step of mesh-to-B-rep conversion particularly consequential: indeed, fitted surface patches must be tangent-continuous across patch boundaries to reproduce the smooth surface character of the intended design [23].

The development of AI generative models is oriented towards the direct production parametric CAD geometry that can bypass the mesh-to-B-rep. Willis et al. [16] pointed out how autoregressive transformer models are capable of generating sequences of operations that, when executed, produce valid B-rep geometry that can be directly editable within standard CAD environments. Indeed, they are able to allow parametric modification by designers and engineering without any conversion step. Nevertheless, there are some current limitations that are an active area of research like the ones regarding restricted geometric complexity, sensitivity to out-of-distribution construction patterns, etc.

2.7.3 Craft, Finishing, and the Limits of Digital Fabrication

Craft finishing and manual fabrication are also some aspects of the AI-to-manufacture gap at the object scale that, even if not fundamental at the engineering component scale, play a fundamental role in the quality of the final objects. Many high-quality objects—such as furniture in solid wood, ceramics, and glass—require a set of manual finishing processes—e.g., sanding, stitching, fire-polishing—that cannot extract and are controlled from digital fabrication files and that AI-generated geometries cannot capture or reproduce.

Some researchers, like Menges and Ahlquist [24], have started exploring the use of AI tools to support crafting intelligence and practitioners by providing structural feedback, optimization of material allocation, or generating form variants for manual evaluation, while still not displacing the craft process itself. Compared to the generation or optimization modalities, which need a degree of fabrication precision and geometric purity that craft processes cannot reliably provide, this supporting role—which exemplifies the augmentation modality—may prove more fruitful in craft contexts.

2.8 Benefits, Limitations, and Open Research Directions

2.8.1 Demonstrated Benefits

The implementation of AI-assisted procedures in object design has shown an undeniable support to well-established practices and several benefits in both industrial and academic studies. Concerning structural performance, neural surrogate models and topology optimization have maintained or improved mechanical performance in object design [5, 7], while reducing object structural systems—such as furniture frames, product enclosures, bicycle components—of a percentage of around 30 to 60%.

In design exploration, VAE-based latent space navigation has enlarged the exploration territory in regards to early-stage design ideas and conceptualization and surface typological configurations [11] with a special focus on footwear and furniture design. In commercial application, the parametric mass-customization systems—like the one proposed by Manavis et al. [20]—could definitely benefit from the implemented of AI-assisted parametric customization that could sensibly help in producing solutions for individual customers at a reasonable time and cost rate. Finally, in integrated design workflows, the combination through sequential multimodality between diffusion models-based ideation, latent space form navigation, and fabrication-constrained topology optimization has shown the possibility to produce designs with performance characteristics that would be hardly achievable through conventional processes, especially in terms of objects where the combination between aesthetic form and structural efficiency is extremely meaningful.

2.8.2 *Principal Limitations*

One of the most recognized limitations at the object scale is the possibility to properly formalize aesthetic judgements within AI optimization objectives. Existing AI models perform very well in optimization for structural coherence, manufacturing costs, and material efficiency, but can hardly produce optimization in regards to aesthetic coherence, cultural legibility, and inner qualities that mark a difference between geometrically interesting shapes and meaningful design objects. This limitation is still intrinsically present in contemporary AI generative systems and defines the undeniable contributions that the human counterpart still can bring to AI-assisted object design workflows [25]. Moreover, the representation gap is still a big challenge, considering that the conversion AI-generated point cloud or mesh geometries to fabrication-ready items still requires a fundamental contribution from designers both for VAE and diffusion models-based generations [12, 16, 26, 27]. A third limitation is the training data challenge—the difficulty to gather large and well-curated datasets of design objects in three-dimensional form—especially for specific design traditions where archival procedures are limited or not considered fundamental [28].

2.8.3 *Open Research Directions*

Summarizing the critical insights from Sects. 8.2 and 8.3, it is possible to identify some specific research directions that can be of particular importance for AI-assisted object design. First is the development of performance-informed generative models that can combine together the optimization of structural performance and the production of aesthetically coherent forms. To achieve this objective, it is required to implement differentiable performance evaluation methods for complex object geometries and to integrate aesthetic coherence metrics in multi-objective training frameworks [17].

Second is the need to develop AI systems that can understand materials as a proper design medium—rather that only a performance parameter—and understand how material properties can shape themselves as design possibilities and open up the process to a different family of forms [2].

Third is the research on AI-supported interfaces for craft practitioners such as models that can improve craft practices without displacing it and that can, at the same time, provide structural feedback, material optimization, and flexible form generation, while maintaining manual fabrication at the core of design activity due to its technical and cultural significance [24]. This last point represents the concrete possibility to extend and augment knowledge in craft fabrication techniques while providing AI-supportive tools simultaneously addressing some other peripheral problematics.

2.9 Conclusion

This chapter has examined AI-assisted design and digital prototyping at the scale of the designed object—chairs, shoes, luminaires, furniture, wearables, and the wider domain of everyday artefacts—and explored the implementation of topology optimization, neural surrogate models, generative adversarial networks, variational autoencoders, diffusion models, and parametric CAD systems in the domains of structural performance optimization, aesthetic form generation, and mass customization contexts. What emerged is that, at the object scale, there are still some distinctive challenges that AI integration must face: the need to link together structural features and optimization with aesthetic form coherence; the role of material in object character and identity; and the consciousness that designed objects not only have to satisfy structural and manufacturing constraints but also have to communicate and deliver cultural meaning while supporting embodied use.

An interesting pattern is revealed from all the examples drawn from the literature (Laarman's topology-optimized furniture [6], Umetani's VAE-based car form exploration [11], Salles and Gyi's anthropometrically personalized insoles [3], and Manavis et al.'s parametric furniture mass-customization system [20]): AI-assisted object design it is truly effective if it works as a collaboration between different AI systems, each one addressing a specific problem within the design-to-fabrication process, and a human intelligent counterpart who provides the cultural knowledge and evaluative judgement that current AI models cannot still synthetize.

The subject of Chap. 3—the transition from the object scale to the building and large-scale artefact scale—requires a paradigmatic change in the field of design problems that goes beyond a simple change in output scales. If object design problems are dimensionally and ergonomically constrained, intimate and explorable by a single designer or a small team of professionals, architectural ones are public, involve multiple stakeholders, legally and institutionally regulated, and contain within themselves a physical, cultural, and social context that lasts on a temporal scale incredibly more if compared to any designed object. For this reason, AI methods, workflows, and human–machine collaboration procedures that were found to be productive at the scale described in this chapter require an ontological reconfiguration to address that problematics and challenges presented in Chap. 3.

References

1. Oxman R (2017) Thinking difference: theories and models of parametric design thinking. Des Stud 52:4–39. https://doi.org/10.1016/j.destud.2017.06.001
2. Ashby MF, Johnson K (2013) Materials and design: the art and science of material selection in product design, 3rd edn. Butterworth-Heinemann, Oxford
3. Salles AS, Gyi DE (2012) The specification of personalised insoles using additive manufacturing. Work 41(S1):1771–1774
4. Cootes TF, Taylor CJ, Cooper DH, Graham J (1995) Active shape models—their training and application. Comput Vis Image Underst 61(1):38–59. https://doi.org/10.1006/cviu.1995.1004
5. Sigmund O (2001) A 99 line topology optimization code written in Matlab. Struct Multidiscip Optim 21(2):120–127. https://doi.org/10.1007/s001580050176
6. Bollinger K, Grohmann M, Tessmann O (2010) Structured becoming: evolutionary processes in design engineering. Archit Des 80(4):34–39
7. Inc A (2020) Generative design in fusion 360: technical overview. Autodesk white paper, San Francisco
8. Nie Z, Jiang H, Kara LB (2020) Stress field prediction in cantilevered structures using convolutional neural networks. J Comput Inf Sci Eng 20(1):011002. https://doi.org/10.1115/1.4044097
9. Chang AX, Funkhouser T, Guibas L, et al (2015) ShapeNet: an information-rich 3D model repository. arXiv:1512.03012
10. Mirza M, Osindero S (2014) Conditional generative adversarial nets. arXiv:1411.1784
11. Umetani N (2017) Exploring generative 3D shapes using autoencoder networks. In: SIGGRAPH Asia 2017 technical briefs, pp 1–4
12. Regenwetter L, Curry B, Ahmed F (2022) Deep generative models in engineering design: a review. J Mech Des 144(7):071703. https://doi.org/10.1115/1.4053859
13. Vahdat A, Williams F, Gojcic Z, Litany O, Fidler S, Kreis K (2022) Lion: latent point diffusion models for 3d shape generation. Adv Neural Inf Process Syst 35:10021–10039
14. Jun H, Nichol A (2023) Shap-E: generating conditional 3D implicit functions. arXiv:2305.02463
15. Cross N (2023) Design thinking: Understanding how designers think and work. Bloomsbury Publishing
16. Willis KD, Jayaraman PK, Lambourne JG, Chu H, Pu Y (2021) Engineering sketch generation for computer-aided design. In Proceedings of the IEEE/CVF conference on computer vision and pattern recognition, pp 2105–2114
17. Vlassis N, Ma R, Sun W (2020) Geometric deep learning for computational mechanics part I: Anisotropic hyperelasticity. Comput Methods Appl Mech Eng 371:113299. https://doi.org/10.1016/j.cma.2020.113299
18. Raissi M, Perdikaris P, Karniadakis GE (2019) Physics-informed neural networks: a deep learning framework for solving forward and inverse problems involving nonlinear partial differential equations. J Comput Phys 378:686–707. https://doi.org/10.1016/j.jcp.2018.10.045
19. Kyratsis P, Manavis A, Davim JP (Eds) (2023) Computational design and digital manufacturing. Springer

20. Manavis A, Minaoglou P, Efkolidis N, Kyratsis P (2024) Digital customization for product design and manufacturing: a case study within the furniture industry. Electronics 13(13):2483. https://doi.org/10.3390/electronics13132483
21. Karras T, Laine S, Aila T (2019) A style-based generator architecture for generative adversarial networks. In: Proceedings of the IEEE/CVF conference on computer vision and pattern recognition (CVPR), Long Beach, pp 4401–4410. https://doi.org/10.1109/CVPR.2019.00453
22. Sbrugnera Sotomayor NA, Caiazzo F, Alfieri V (2021) Enhancing design for additive manufacturing workflow: optimization, design and simulation tools. Appl Sci 11(14):6628. https://doi.org/10.3390/app11146628
23. Pietroni N, Campen M, Sheffer A, Cherchi G, Bommes D, Gao X, ... Livesu M (2022) Hex-mesh generation and processing: a survey. ACM Trans Graph 42(2):1–44
24. Menges A, Ahlquist S (2011) Computational design thinking: computation design thinking. John Wiley & Sons
25. Lawson B (2006) How designers think
26. Perna V (2025) Technicity between Prediction and autonomy: Ceci n'est pas une chaise (This is not a chair). In: Digital product design and manufacturing, pp 179–197. CRC Press. https://doi.org/10.1201/9781003518198
27. Perna V, Kyratsis P (2025) Diffusion models in product design: optimizing form toward digital prototyping. CAD/CAM: Comput Aided Design Manuf 17:21. https://doi.org/10.1515/9783111585062
28. Bommasani R, Hudson DA, Adeli E, Altman R, Arora S, von Arx S, ... Liang P (2021) On the opportunities and risks of foundation models. arXiv preprint arXiv:2108.07258

Chapter 3
The MEDIUM(S): Diffusion Models and GAN for Building Scales and Single Artefacts

Abstract The previous chapter served to define the theoretical framework of AI-assisted design and discuss its possible application in regard to the scale of the individual object—e.g., chairs, shoes, lamps, and more generic consumer artefacts. The architectural scale is now the objective of the present chapter whether we refer to the dimension of the building, the large-scale artefact, or the interior design one. In this scale, the transition is not just scalar but rather qualitative: design problems related to architecture are open instead of geometrically closed; multi-stakeholder instead of individually directed; subject to constraints that simultaneously can be programmatic, structural, environmental, etc. while being embedded in complex physical and cultural contexts that might resist to the reduction that characterizes computable objectives. The present chapter investigates the way in which generative adversarial networks (GANs), diffusion models, convolutional neural networks, and graph-based generative architectures can be applied to address the four main architectural design challenges and tasks at the building scale: facade generation and style synthesis, floor plan layout and spatial program generation, structural form-finding, and building performance optimization. Furthermore, this chapter discusses these AI technical abilities in relation to the intellectual context of architectural design and intersects issues concerning authorship, creativity, and novel aesthetics that AI-based procedures can foster. The discussion is grounded in three case studies drawn from the existing literature: Chaillou's ArchiGAN stack for apartment building generation, Nauata et al.'s House-GAN system for graph-conditioned floor plan synthesis, and del Campo's Deep House project, where estrangement and defamiliarization are the theoretical lens through which GAN-generated residential form is addressed. The conclusive part presents the benefits, limitations, and the specific research challenges between these AI practices at the architectural scale.

Keywords Architectural AI · Generative adversarial networks · Diffusion models · Floor plan generation · Facade synthesis · Building performance optimization · Neural architecture · Design creativity · Estrangement · Parametric design · BIM · Structural form-finding

P. Kyratsis et al., *AI Models for Design and Digital Prototyping*,
Manufacturing and Surface Engineering, https://doi.org/10.1007/978-3-032-35144-9_3

3.1 Introduction

The scale hierarchy proposed in Chap. 1 of this volume—from the individual object through the building to the urban territory—is not just a simple spatial classification. It represents a growing progression in the character of design problems and the challenges related to them: their degree of geometric openness, the heterogeneity and multiplicity of constraints and stakeholders they involve, the temporal scales on which their consequences are project, and the cultural and social dimensions they intersect and engage with. Furthermore, the transition from the object scale examined in Chap. 2 to the architectural scale discussed in this section can be considered a groundbreaking point of the scalar transitions in this book: indeed, it crosses the threshold where design is not anymore an monodirectional relationship between a designer and a user but transforms into a complex system of relationships among a designer, an institution, a site, a community, and a cultural, social, and geographical, and tradition accumulated over millennia of building.

At this point, it is easy to understand why architecture can be intended as the original design discipline: a specific domain where, over the centuries, the full complexity of the human-made environment has been synthetized, theorized, and negotiated. The intellectual tradition behind architecture is undoubtedly of immense richness: spanning from the Vitruvius's triad based on firmitas, utilitas, and venustas, to the Enlightenment's codification of building through the concept of typology, until arriving at the Industrial Revolution of the twentieth century and successive revolutions in structural technology, spatial concept, and cultural reference. The emergence of AI in architectural design must be understood in the light of this background and as a way not to replace the accumulated intelligence that lies behind architectural culture, but rather as a new instrument that interacts and relates with that through novel and still to be completely discovered operative categories.

In regard to the architectural scale, if the most diffused AI model families—generative adversarial networks, diffusion models, graph-conditioned generative architectures, and transformer-based spatial models—were already analysed in Chap. 1, nevertheless, in this one, the interest focused on their application in architecture design to understand the models' capabilities when they address some specific demands of the discipline: the generation of structurally coherent, climatically adequate, culturally grounded, and formally inventive, building facades; the calculation for floor plan layouts that can fulfil circulation logic, programmatic requirements, structural uniformity, and spatial quality; the form-finding calculation of structural systems that might be efficient, buildable, and architecturally expressive; and the optimization of non-linear interrelated building performances such as energy, daylight, acoustic, and occupant comfort dimensions.

The discussion is enriched by three exemplary case studies selected from the recent literature and that help to ground this theoretical and technical analysis in contemporary architectural practice. Chaillou's ArchiGAN [1], the results of a Harvard GSD thesis, shows a three-model generative adversarial networks (GAN) pipeline that can generate apartment building floor plans starting from the data of an urban parcel through an iterative and sequential process based on footprint massing. Nauata et al.'s House-GAN [2] uses graph-conditioned GAN generation to define architectural program as a relational graph of rooms and adjacencies where GAN generates alternative possibilities that could satisfy these relations. Del Campo's Deep House project [3] represents an applied GAN-based generative method where the produced architectural imagery is grounded on the theoretical concept of estrangement—specifically, the production of family of forms that are concurrently recognizable and alien, familiar and uncanny, etc.—and intersects a broader discussion related to creativity, novelty, and the relationship between human and artificial intelligence in architectural production.

This chapter is organized as follows. Section 3.2 defines architectural scale as a proper design domain, and identifies the characteristics through which it differs from the object scale and, at the same, suitable AI strategies. Section 3.3 explores AI applications to the specific fields of façade generation and architectural style synthesis. Section 3.4 investigates AI-based floor plan and spatial layout generation, where the ArchiGAN and House-GAN case studies are presented. Section 3.5 focuses on structural system production and structural form-finding, while Sect. 3.6 intersects the topics of architectural creativity, authorship, and aesthetic in the age of AI presented del Campo's theoretical framework and Deep House project. Section 3.7 examines the possible integration between Building Information Modelling (BIM) and AI not only in terms of generated forms but also buildable design documentation. Section 3.8 discusses benefits, limitations, and open research directions. Finally, Sect. 3.9 is a conclusive bridge to Chap. 4 and its further investigation across the urban scale.

3.2 The Architectural Scale: Characteristics and Design Challenges

A necessary precondition to evaluate the impact of AI methods at the scale of architecture is to understand the peculiarities of the design problems related to that and the implications that they have in terms of AI integration. Indeed, the architectural scale is defined by a series of properties that simultaneously determine the potential contributions and limitations that AI currently faces.

3.2.1 Openness, Multi-Constraint, and the Problem of the Brief

If the object design problems were normally considered as geometrically closed, measurable in terms of performance, and directly deployed to a user-testing phase, on the contrary architectural design problems have an inner openness that conflicts when it comes to their computational formalization. In architecture, a design brief is not a complete specification of the final desired product but more an ambiguous, partial, and evolving system of requirements that the designer/architect should consider to produce a design hypothesis. For example, a brief for a residential building can deal with the number of rooms, their areas and adjacencies, but cannot define the final three-dimensional spatial experience of the house; a cultural building brief can be really thorough in relation to program, site constraints, and budget, but rarely giving an exhausting explanation on the relationship between public and private spaces; the controlling of natural light; or the architectural character that the building should have in its urban context.

Therefore, openness in architectural design means that the final results cannot be reduced only to constraint satisfaction or objective optimization like it happens for many engineering design problems. An architect not only in unison must address structural, programmatic, performance, regulatory planning, and client requirements, but also all the implicit ones connected to architectural quality such as material character, cultural references, spatial experience, and tectonic expression. All of these last categories are not specific in any brief but their satisfaction is expected to happen from the architectural community as a sign of professional competence. AI models that tend to address one of these requirements in isolation risk to perform an optimization related to just a partial problem that misses those that are the real central requirements that characterize architectural quality [4].

3.2.2 Context, Site, and the Situated Character of Architecture

Architecture is irreducibly situated: every building occupies a specific site, situated in a specific urban context, submerged in a specific and intertwined cultural and historical layer. Therefore, the design intelligence that activates to produce a building in the historic centre of Vienna qualitatively differs from the one needed for the same functional program happening on a rural site in Brazil or a suburban area in Seoul. This can be explained not only in relation to the differences in climate and structural requirements among diverse countries but, above all, considering the relations that the buildings must weave with its cultural and historical context—the traditions that it decides to challenge or continue; the references it acknowledges—and that represent a primary dimension in architectural meaning.

Existing AI models, even though trained on large datasets that collect examples from diverse contexts, lack satisfactory description of the contextual specificity for which each of those examples can be considered architecturally appropriated or not in a specific setting. To be more specific, if a GAN is trained on a corpus of European apartment buildings, surely it will learn about the regularities of that building typology but not about the contextual intelligence—the ability to read a specific urban block, street section, or a specific historic layer of the built environment—that an experienced professional architect always brings as an argument in every design situation [5]. This very contextual understanding limitation is one of the main constraints related to the applicability of AI generative models in architecture, and an aspect of the discipline that should be seriously considered and further explored since AI tools are already widely integrated into architectural practice.

3.2.3 Program, Typology, and Spatial Logic

A fundamental tool in architectural design intelligence is the concept of typology, intended as the classification of buildings according to their program, spatial organization, structural system, etc. A good typological knowledge allows architecture to design and draw on accumulated design solutions for repeated building programs (e.g., a hospital, a school, an apartment block, and a museum) while these known solutions are applied to the specific conditions of each project and building area on which it will be located.

AI models for architectural design relate to typological knowledge mainly throughout training on datasets of built examples, where they extrapolate statistical regularities in the spatial organization of floor plans, the distribution of programmatic elements, and the relationships between structural system and spatial organization.

This typological learning is a genuine and practically significant capability: it enables AI systems to generate floor plan configurations that respect the implicit spatial logic of a given program—the requirement that bedrooms have access to natural light, that service spaces are positioned along party walls, and that public circulation is separated from private areas. Nevertheless, current AI typological models are not able to capture the creative renegotiation of typological conventions that characterizes the most significant architectural work—the reimagining of the hospital as a therapeutic landscape, the reinvention of the museum as a public space, and the transformation of the apartment block into a social condenser. This creative typological intelligence, which distinguishes great architecture from competent building, remains a challenge beyond the current reach of AI systems [6].

3.3 Facade Generation and Architectural Style Synthesis

In architecture, the façade is the primary mediating surface and connection that a building has between the interior space and the urban context and the dimension of architectural design which is most accessible to AI generative models. Indeed, differently from a three-dimensional spatial organization of a floor plan, that needs complex and well-established relation structures among circulation, rooms, structure, and program, the façade can be intended as two-dimensional image composition characterized by elements such as material, texture, fenestration rhythm, proportion, and ornamental articulation, that can be easily extracted from a training datasets based on photographs or drawing through the use of image-based convolutional and generative architectures.

3.3.1 *Convolutional Approaches to Facade Analysis and Generation*

The automatic segmentation and classification of facade elements (windows, doors, balconies, pilasters, cornices) from photographs has been reached with increasing accuracy using convolutional neural networks (CNNs) for facade analysis. These tools permitted the construction of structured facade datasets from large collections of unstructured street photography [7]. Such structured datasets, where facade elements are labelled and positioned, are at the core of the most recent training procedures of conditional GAN models for controlled facade generation: starting from a specific façade composition—in terms of element types and their approximate spatial position—a cGAN can produce photorealistic renderings that are coherent with the composition brief.

Isola et al. developed the pix2pix framework [8]—which uses paired training data and is based on conditional image-to-image translation—that has been widely used in architectural façade generation. The pix2pix model, which is trained on paired datasets of facade label maps and corresponding photographs, can learn to transform abstract spatial compositions—specific blocks of colours for windows, wall, and other architectural elements—into hyper-realistic renderings. The impact of this ability is undoubtedly meaningful in the early stages of the design process since it allows an architect to quickly visualize elements such as materiality and façade rhythm without wasting a huge amount of time spent in modifying the three-dimensional model. Therefore, cGANs' ability to generate coherent renderings in a few seconds instead of hours speeds up the evaluation capacity for design alternatives from the façade development stage.

The well-known style transfer and style synthesis capabilities—namely the application of a specific historical or contemporary architectural style to a façade geometry—are the results of the use of a range of GAN architectures, such as CycleGAN [9] for unpaired style transfer and StyleGAN [10] for the generation of facades with specified stylistic characteristics. The analysis performed by Chaillou on architectural style learning [1] showed that if GAN models are trained on a style-based dataset of floor pan and facades, they are able to develop internal representation and can provide not only aesthetically appealing visual texture but also results grounded on a thorough spatial and compositional logic. More specifically, if a model is trained on Baroque floor plans, it can extract and learn the hierarchical spatial organization of Baroque planning and apply it to new geometries in a way that goes beyond simple imitation.

3.3.2 Diffusion Models for Architectural Visualization

Stable Diffusion™ [11], DALL·E 2™, Midjourney™, and other text-conditioned diffusion models have rapidly caught the interest of architecture professionals and became essential tools for concept visualization and client communication. Their capacity to produce photorealistic architectural images from natural language description empowers architects to explore, visualize, and communication a huge variety of design directions since the early stages of the projects, and without the need for any specific technical development.

This early-stage visualization capability has shown its potentialities in regard to competition submissions, client presentations, and also urban design consultations where—before a design development is enough advanced to support conventional rendering phases—the communication of spatial and material character of a project is fundamental.

Nevertheless, limitations in this category are more than significant and should be fully understood for its responsible use in professional practice. Currently, diffusion models can produce aesthetically appealing images but not coherent three-dimensional forms: a building in a diffusion-generated image can be visually compelling but impossible under a structural perspective, inconsistent between the produced views, or even spatially incoherent from an architectural perspective. Rombach et al. [11] explained this limitation in their analysis of latent diffusion model capabilities: it resides in the inner nature of image diffusion models that are basically statistic image generations and, therefore, not proper architectural design systems, and mostly tools for architectural visualization rather than architectural generation.

However, some architecture-specific fine-tuned diffusion models, trained on curated datasets of high-quality architectural photography and drawings, can be considered a step-ahead in architectural coherence if compared to general-purpose models [12]. An active research direction that may reduce the gap between architectural visualization and architectural generation is represented by diffusion models trained on specific architectural constraints such as site boundaries, program specification, and urban context images. Nevertheless, as Rombach et al. also highlighted domain-specific fine-tuning definitely can improve the architectural quality of a set of produced images, even though the absence of three-dimensional spatial reasoning is one the most worrying phenomenon in this field and persists across almost all the existing image-based diffusion architectures.

3.4 Floor Plan Generation and Spatial Layout

The generation of floor plans—the horizontal projection of a building's spatial organization—is currently the architectural AI application that benefitted from an intense level of research attention and that can produce results among the most technical ones.

A floor plan embeds the fundamental spatial logic of a building: the number, size, and shape of rooms; their programmatic identity; their mutual adjacency and separation; their relationship to circulation, structure, and envelope. The capacity to generate floor plans—that can respect both programmatic and spatial requirements—has to be considered of interest especially for all those early-stage design explorations and design automation procedures when typological conventions are well-defined and established.

3.4.1 Image-Based GAN Approaches: The ArchiGAN Pipeline

Chaillou's ArchiGAN [1], developed as a Harvard Graduate School of Design thesis and published in 2019, represents one of the most comprehensive early demonstrations of GAN-based floor plan generation at the building scale. The system comprised a cascaded pipeline of three GAN models, each addressing a distinct level of the design hierarchy: the first model generated building footprint geometry from urban parcel data derived from GIS sources for the city of Boston; the second model generated room layout and fenestration from the building footprint and a designer-specified entrance position; and the third model generated furniture layout from the room configuration. By nesting these three models sequentially—with human design input at each transition point—Chaillou created a generative workflow capable of producing complete apartment building designs from urban site data, with the architect exercising judgement at the transition between each generative stage.

A particularly significant finding of the ArchiGAN work was the demonstration that GAN models trained on style-specific datasets of floor plans learn the spatial logic—the mechanics of space, in Chaillou's terminology—characteristic of those styles, not merely their visual appearance. A model trained on Baroque floor plans generates layouts organized around enfilade sequences and hierarchical room progressions; a model trained on Manhattan apartment plans generates layouts organized around corridor circulation and compact service groupings. This style-conditioned spatial logic is not explicitly programmed but emerges from the statistical structure of the training data, suggesting that GANs can learn architectural intelligence from architectural examples without being given explicit rules. Chaillou's framework—using image representations of floor plans as the data format for GAN training and inference—established a methodology subsequently adopted and extended by numerous research groups working on architectural floor plan generation [2, 13].

3.4.2 *Graph-Conditioned Generation: The House-GAN Approach*

One of the fundamental limitations of the application of image-based GAN in architecture is their difficulty to fulfil specific programmatic requirements. For example, in the ArchiGAN framework, the designer can control the generation selecting a style-training corpus of data and the position of the entrance doors and windows, but does not have agency in the number of room, the spatial requirement for a study, or the best adjacency factor between a dining space and the kitchen. This limitation has been addressed by the House-GAN framework, proposed by Nauata et al. [2] and its extension House-GAN++ [14], where the GAN is conditioned on an explicit graph representation of the architectural program and programmatic requirements.

Using the House-GAN architecture, an architect can define the program using a bubble diagram: a node-based graph representation where rooms are labelled by type (bedroom, living room, bathroom, etc.) and the edges require adjacencies between them. The program graph is considered the conditioning input to a relational graph convolutional network that generates a floor plan based on adjacencies in a bounding box geometry. This system provided strong performance when tested on a dataset of real estate floor plans extracted from the RPLAN dataset [15], and generated configurations that were satisfying in terms of adjacency relations and that comparable to other floor plans created by alternative automated methods. In some cases, the results from House-GAN were also exceeding in some metrics the ones from the compared models.

Therefore, House-GAN can be considered a significant methodological advancement from image-based approaches because it involves architects' programmatic agency: the professional decides what the space must contain and how those things should relate between each other, and the AI created spatial outcomes that can fulfil these requirements. The extension in House-GAN++ [14] to larger, more complex floor plans with improved room shape quality and tighter adjacency constraint satisfaction represents progressive refinement towards architectural deployability, though the system's current limitation to single-floor residential layouts constrains its scope of application.

3.5 Structural Form-Finding at the Building Scale

It is intrinsically wrong to consider the structural system of a building as a mere engineering necessity: in reality, it is the condensation of a tectonic tradition and one of the primary vectors of architectural meaning. Whether we refer to the exposed concrete frame of Le Corbusier, the cable net of Frei Otto, the branching steel columns of Calatrava's railway stations, the timber lattice shell of Shigeru Ban, etc., in each case the structural system is the natural continuation of the architectural and spatial idea behind the building rather than just a physical support. Therefore, implementing AI to structural form-finding at the building scale means engaging not only an engineering optimization but also the architectural expression of it, where the two different cannot be divided from one another.

3.5.1 AI-Assisted Structural Topology at the Building Scale

In Chap. 2, different topology optimization methods were introduced at the object scale. At the building one, considering the problem scale, the diversity of loading conditions, and the architectural constraints on structural expression, a qualitative adaption to these methods needs to be done. In architecture, structural topology optimization has been applied to the design of long-span roof structures, transfer structures spanning irregular column grids, and facade bracing systems, with the aim of generating structural configurations that concentrate the material along the principal load path and using formal innovations that go beyond traditional structural solutions [16].

The real computational challenge of this category—influenced by the building geometry itself, the multiple simultaneous loading conditions (gravity, wind, seismic, thermal), and the regulatory performance requirements (fire resistance, vibration limits, deflection criteria, etc.)—has pushed forward the development of neural network surrogates that could perform structural analysis at a lower cost if compared to conventional finite element analyses [17]. Such surrogate approaches opened the possibility to explore structural configurations that would be conceptually inaccessible if using direct analysis-based methods, and enabled a design process where a rich space of structurally informed alternatives is given to the architect instead of a single structural concept.

3.5.2 Machine Learning for Structural Performance Prediction

Machine learning systems have not been used only to topology optimization, but also for the prediction of structural performance metrics (e.g., inter-storey drift under seismic loading, natural frequency under wind-induced excitation, buckling load for slender structural elements), and the geometric and material description of structural systems.

So-called regression neural networks—trained on databases of structural analyses—have shown a sufficient grade of accuracy for early-stage design screening [18], which is the ability to extract, from a large population of structural configurations, the one that could potentially fulfil the needed performance requirements and before committing to expensive full structural analysis.

The integration of structural performance in generative design loops, through the use of predicted performance metrics such as objectives or constraints inside a Bayesian optimization or generic algorithms, has shown promising outcomes for the design of diagrid facades [17], moment-resisting frames, or grid shell structures [19]. These workflows enhance the architects' capacity to explore simultaneously structural efficiency, material economy, and architectural aesthetic: a combination of inputs and outputs that traditional sequential *if–then* methodologies could not perform. An architect that manages to control these AI-based tools can receive continuous structural feedback as soon as a formal decision is made, and achieve an integrated structural and aesthetic outcome that conventional practices—where the engineering part is often engaged after the formal definition of the building—rarely could achieve.

3.6 Creativity, Authorship, and Aesthetic Novelty in AI-Generated Architecture

Undoubtedly, creativity is the dimension of AI's engagement in architecture that raises the more fundamental and ethical questions. Architecture has always been intended as a creative discipline where the designer's individual intelligence, cultural formation, and artistic sensibility are conveyed and expressed through built form. The appearance of AI systems capable of generating architectural imagery, floor plan configurations, and structural forms that are comparable to the ones produced by human intellect poses again the question, of what creativity in architecture consists of, and if AI can have a role in understanding, replicating, and extending it.

3.6.1 Boden's Creativity Typology and AI-Generated Architecture

In her 'Creativity and Artificial Intelligence' [20], Margaret Boden identifies different types of creativity: combinational creativity (the novel interpolation and combination of already existing ideas); exploratory creativity (the exploration of an existing contextual space); and transformational creativity (the restructuring of that contextual space). Her classification serves a useful lens and analytical framework to understand the role of AI in architectural generation. All the current AI systems—e.g., GANs that interpolate between training examples, diffusion models that synthesize new images from learnt statistical distributions, VAEs that navigate latent morphological spaces—can be mainly classified as combinatory or exploratory according to Boden's classification: indeed, they perform recombination and exploration of the architectural forms inserted in their training data, and they produce outputs that are novel instances within that space rather than a transformation of the space itself.

In the analysis of his Deep House project [3], Matias del Campo reflects on the relationship between AI-generated images created by his firm SPAN and the concept of new in the architectural discipline. Starting from Boden's analysis and the idea that the human brain works as a kind of 'giant remixer'—that recombines already known concepts into surprising configuration that are interpreted as 'creative'—he argues that the production of neural networks in architecture have to be intended as 'estranged objects': forms and shapes that simultaneously appear as architecturally recognizable but formally alien, since they are the product of a statistical learning process that interpolates existing architectural knowledge in a way no design would do. This process—the estrangement/defamiliarization of the known—according to del Campo can be considered a proper creative dimension of AI-created forms, even if it practically stands far from the transformational creativity described by Boden.

3.6.2 The Deep House: Estrangement as Design Method

The Deep House is a project by Matias del Campo and Sandra Manninger (founders of the studio SPAN Architecture)—documented in papers and books [3]—and that provides some of the most theoretically elaborated contributions to the discussion on contemporary GAN-based architectural design practices.

The project was designed for a neuroscientist in Austria that requested himself a house where AI-assisted design methods could be tested to produce something 'different' and that could differ from the canonical architectural conventions related to residential architecture.

For the design process a wide range of artificial neural networks is used—spanning from GAN to CNN architectures—with the aim of generating architectural outcomes that were implanted in the design process as a corollary of formal suggestions and spatial ideas, rather than a direct design outcome. Therefore, the AI images were intended as a generative stimulus: they were used to produce formal configurations that SPAN could not have imagined independently and that combined several architectural fragments from the training dataset—roof forms, aperture arrangements, material textures, spatial transitions—to generate what Surrealists would call an 'exquisite corpse', a formally coherent agglomeration at each local point but that would globally result as estranged, unfamiliar, and immune to tradition typological classifications.

For del Campo, this estrangement is not a failure of AI in generating 'correct' architectures, but rather a productive capacity that could help architecture to escape the risk of reproducing the familiar. Therefore, the concept of estrangement—defined as Verfremdung in Brecht's theatrical theory—is used to describe a sort of AI productive alienation that makes the familiar strange and forces architects and viewers to relate with architectural space without the conventional recognized categories forced by typological approaches. Under this perspective, AI is considered not as a generator for architecture but rather a collaborator that introduces a formal otherness in the design process that has to be considered intrinsically creative because it is not subjected to the architect's will, cultural background, and conscious control [3].

The importance of the Deep House in this book is marked by the fact that the project is not only a demonstration of GAN application at the residential building scale but also because it is an attempt to theoretically address AI-generated architectural form also under a professional perspective. The work of Matias del Campo and Sara Manninger shows that integration of AI in architecture is more an intellectual challenge rather than a technical one: it pushes architects to redefine their understanding of creativity, authorship, and relation between machine-generated form and human intelligence, and highlights a distinction between the use of AI as a mere productive tool or as an investigation tool for the generation of a new aesthetic.

3.6.3 Authorship, Attribution, and the Ethics of AI-Assisted Design

The problem of authorship is definitely one of the most delicate questions regarding AI-assisted architectural design and, in the last years, attracted a wide interested from professional bodies and legal scholars. When AI systems are the ones delegated to the production of spatial configurations, formal proposals, and structural solutions that the human architect evaluates, selects, and develops, the conventional role of the architect as the originator of the design is shaken to its foundations, and we are in front of a product of a human–machine collaboration whose implications and legal boundaries have still to be theorized [26].

Some professional bodies like the American Institute of Architects (AIA) and the Royal Institute of British Architects (RIBA) have produced some guidelines for the usage of AI in architectural practice that aims to regulate its involvement in design processes, the attribution of design decisions to human or machine agents, and the professional liability implications of AI-assisted design. Nevertheless, all these proposals are at an early stage and still have to be refined accordingly also to what will happen in terms of future AI integrations in architecture. For this book, the question of authorship is more than a legal formality: the possibility to assess to what extent a human architect is the author of an AI-assisted architecture is useful to determine how much a design expresses a human architectural intention and, therefore, how much that specific outcome is situated in the tradition of architecture as an exclusive humanistic discipline.

3.7 AI Integration with Building Information Modelling

Building Information Modelling (BIM) is the integrated digital representation of building geometry, material specification, structural system, building services, and program information within a single, consistent data model and currently represents the primary digital infrastructure of contemporary professional architectural practice. The integration of AI-generated design outputs with BIM workflows is therefore an important milestone for the practical deployment of AI in professional architecture, but it remains at the same time one of the domains in which the gap between AI research and professional practice reality are currently very far from one another.

3.7.1 The BIM Representation Gap

The outputs of AI generative models for architecture—whether we refer to floor plan images from GAN models, three-dimensional mesh geometries from diffusion models, or point cloud configurations from three-dimensional shape models—are not directly compatible with the BIM data structures used by professional architectural practice. All the most diffused BIM platforms such as Autodesk™ Revit™, Graphisoft™ Archicad™, and Bentley™ AECOsim interpret building geometries as an assembly of families of typed building elements—walls, floors, roofs, columns, beams—to which is associated parameter data: material specification and performance, structural performance characteristics, thermal properties, and cost information. Before being used into BIM environments, AI-generated geometry—which often do not show element typing, parametric associativity, and performance data—have to converted into formats and elements that can be modified and interpreted by the software [27].

Similar to the AI-to-manufacture gap discussed in Chap. 2, the BIM representation gap poses analogous challenges since the manual conversion of AI geometries into BIM models is intensive under a labour perspective and also has the tendency to produce errors. The results are a series of models that lack the parametric associativity and data implementation and richness that give to Building Information Modelling its productivity advantages. Research to produce automated AI-to-BIM conversions—that can use machine learning to classify AI forms by building element type and instil in them parametric properties—are still at an early development stage and still far from what professional practice requires.

3.7.2 AI-Native Architectural Modelling Approaches

Some architectural research groups are developing some AI-native approaches that can bypass the current limitations and generate outputs that can be directly implanted into a BIM environment, in commercial and academic contexts to find systems that generate Revit™ parametric families starting from AI design, or that can produce Grasshopper™ definitions similar to the CAD construction sequences for object design. Nevertheless, these approaches face exactly the same challenges that were discussed at the object scale—restricted geometric complexity, sensitivity to out-of-distribution design patterns—which are even more significant at the building-scale geometry.

The development of large-scale datasets of annotated architectural designs in BIM-compatible formats—a challenge similar to the ShapeNet™ dataset for object geometry [21]—is a fundamental condition for the training of AI models that generate architecturally valid, BIM-compatible outputs. The assembly of such datasets faces significant challenges: architectural BIM models are proprietary, heterogeneous in their data structures and modelling conventions, and subject to intellectual property protections that limit their use for AI training. Industry-academic partnerships and the development of open-source BIM dataset initiatives are the most plausible routes towards the data infrastructure that AI architectural generation at the professional scale requires.

3.8 Benefits, Limitations, and Open Research Directions

3.8.1 Demonstrated Benefits

Several categories of benefits have been recorded in regard to the implementation of AI-generative processes at the architectural scale. GAN-based and diffusion-based visualization tools have enhanced the range of formal explorations and directions in early-stage concept development, and allowed architects to investigate a broad territory of architectural possibilities within a shorter and more compressed timeframes. Chaillou's ArchiGAN [1] showed how a cascaded GAN pipeline has the strength of generating complete apartment building designs interpolating urban parcel data, and introduced a generative workflow where major decision points are set by the architects while the AI is delegated to produce related outcomes.

In structural form-finding, AI-assisted topology optimization—where the combination among structural configurations, material efficiency, and structural investigations comes together—has seriously challenged the results of more conventional engineered structural design, with an impact comparable to the one discussed at the object scale in Chap. 2, but operating at the scale of building structure and facade systems.

3.8.2 *Principal Limitations*

Current limitations regarding AI systems at the architectural scale are most consistent than at the object scale, and they reflect the quantitative and qualitative differences between the two domains highlighted in Sect. 3.2. The most fundamental is the contextual limitation—the AI inability to adequately deal with the natural, historical, and urban context of a given space. Furthermore, all the AI systems that are trained on large, and contextually aggregated datasets learn the statistical correspondences of architecture but not yet the contextual specificity that marks the difference between architecture and generic buildings [5]. Not secondarily, the difficulty of defining complex, multidimensional architectural programs as inputs to generative models creates the so-called programmatic limitation that confines the implementation of AI only to building typologies with strong and stratified conventions. A strong advancement regarding this limitation is the House-GAN framework [2] using through graph-conditioned generation, even though the methodology today is yet limited to single-residential floors layout and the extension to more complex building types is still an open research direction. Finally, the BIM representation gap—namely the existing distance between AI-produced geometries and a BIM-compatible data structure needed in professional contexts—still represents a barrier to a market-oriented implements of these technologies that limits the actual impact of AI-based architectural tools in contexts such as research and concept development.

3.8.3 *Open Research Directions*

Within the architectural scale, three are the research directions of particular importance that can be identified. First is the development of context-aware architectural AI—that can deal with the complexity of specific urban, historical, and climatic contextual information of a building site—through the implementation of training data and model architectures that can depict contextual specificities rather than normalize them. Therefore, a promising direction is the one involving multi-modal models that can interpolate building geometry, urban morphology data, historical archive images, and climate-based information.

Second is the realization of architecturally interpretable AI systems—families of generative models where the outputs are accompanied by explanations of the design process that professionals can assess and evaluate. Indeed, the traditional architectural discourse—the articulation of design intentions and their direct spatial, material, and cultural consequences—needs AI systems to produce not only optimized metrics but also clear design rationale. AI researchers and architectural theorists are more engaged in this field of research dedicated to the articulation of AI generative logic within architectural terms [22, 23].

Third, the critical and theoretical interrogation in the field of architectural creativity—as well-explained by Matias del Campo's work on estrangement [3, 24,

25]—is another priority that situates itself in an overlapping dimension between humanistic theories and technical research direction. The capacity of architecture to implement AI tools without losing its humanistic set of values depends of a proper intellectual framework that can clarify to what extent AI contributes to architectural design, where it cannot contribute, and what most productive forms of human-AI collaborations are for the discipline. Currently, this is one of the frameworks under intense research, and its proper clarification will be one of the intellectual projects impacting architectural culture for the decade to come.

3.9 Conclusions

This chapter discussed the application of different families of AI models—from generative adversarial networks to diffusion models; from graph-conditioned generative architectures to neural network surrogate models; and reinforcement learning systems—to architectural design at the building scale. Since the beginning, the architectural scale has been defined as a quantitative and qualitative different domain from the object-based one examined in Chap. 2: more contextually situated, more multi-stakeholder, more open, and more deeply rooted in the humanistic tradition of a discipline that has always negotiated the balance between cultural and contextual meanings and technical necessities.

The three case studies drawn from the existing literature have connected the technical analysis to real architectural practice. Chaillou's ArchiGAN [1] showed the peculiarity of cascaded GAN pipelines to produce residential floor plans from urban site data, and established a human-AI shared operative workflow that preserves architects' agency in terms of structuring decisions. Nauata et al.'s House-GAN [2] introduces graph-conditioned floor plan generation where architectural program is depicted as a relation graph that can allow program-based spatial configuration generation. In a completely different register, del Campo's Deep House [3] exploited the concept of estrangements and situated within a philosophical dimension the topics of AI creativity and architectural novelty, while demonstrating the potentialities of GAN-generated images to introduce productive otherness—in terms of forms—into the design process through the estrangement of familiar architectural conventions with the aim of opening new directions for design explorations.

In Chap. 4, the transition from the building scale to the urban scale takes the design issues mentioned here to a more intricate level. Indeed, at the urban scale, AI must deal not only with the design of a single building but with the emergent properties of complex urban systems which are characterized by a consistent number of patterns concerning density, movement, ecological performance, and social interaction that arise from the collective configuration of many buildings, streets, and public spaces over time. In this perspective, the contextual, multi-stakeholder, and temporal complexity of urban design problems almost makes the architectural challenges present in this current this chapter limited and tractable.

Chapter 4 will examine how AI methods—spanning from urban morphology analysis to generative urban master planning—are moving some initial steps to address this complexity and where the appropriate limits of AI agency in cities design actually stands.

References

1. Chaillou S (2019) Archigan: a generative stack for apartment building design. NVIDIA Corporation
2. Nauata N, Chang K-H, Cheng C-Y, Mori G, Furukawa Y (2020) House-GAN: relational generative adversarial networks for graph-constrained house layout generation. In: Vedaldi A et al (eds) computer vision—ECCV 2020. Springer, Cham, pp 162–177. https://doi.org/10.1007/978-3-030-58452-8_10
3. del Campo M (2022) Deep house—datasets, estrangement, and the problem of the new. Archit Intell 1:12. https://doi.org/10.1007/s44223-022-00013-w
4. Oxman R (2017) Thinking difference: theories and models of parametric design thinking. Des Stud 52:4–39. https://doi.org/10.1016/j.destud.2017.06.001
5. Batty M (2013) The new science of cities. MIT Press, Cambridge MA
6. Cross N (2011) Design thinking: understanding how designers think and work. Berg, Oxford
7. Martinović A, Mathias M, Weissenberg J, Van Gool L (2012) A three-layered approach to facade parsing. In European conference on computer vision, pp 416–429. Berlin, Heidelberg: Springer, Berlin, Heidelberg
8. Isola P, Zhu J-Y, Zhou T, Efros AA (2017) Image-to-image translation with conditional adversarial networks. In: Proceedings of the IEEE conference on computer vision and pattern recognition (CVPR), Honolulu, pp 1125–1134. https://doi.org/10.1109/CVPR.2017.632
9. Zhu J-Y, Park T, Isola P, Efros AA (2017) Unpaired image-to-image translation using cycle-consistent adversarial networks. In: Proceedings of the IEEE international conference on computer vision (ICCV), Venice, pp 2223–2232. https://doi.org/10.1109/ICCV.2017.244
10. Karras T, Laine S, Aila T (2019) A style-based generator architecture for generative adversarial networks. In: Proceedings of the IEEE/CVF conference on computer vision and pattern recognition (CVPR), Long Beach, pp 4401–4410. https://doi.org/10.1109/CVPR.2019.00453
11. Rombach R, Blattmann A, Lorenz D, Esser P, Ommer B (2022) High-resolution image synthesis with latent diffusion models. In: Proceedings of the IEEE/CVF conference on computer vision and pattern recognition (CVPR), New Orleans, pp 10684–10695. https://doi.org/10.1109/CVPR52688.2022.01042
12. Achiam J et al (OpenAI) (2023) GPT-4 technical report. arXiv:2303.08774, https://doi.org/10.48550/arXiv.2303.08774
13. Wu W, Fu X, Tang R et al (2019) Data-driven interior plan generation for residential buildings. ACM Trans Graph 38(6):1–12. https://doi.org/10.1145/3355089.3356556
14. Nauata N, Hosseini S, Chang KH, Chu H, Cheng CY, Furukawa Y (2021) House-gan++: Generative adversarial layout refinement network towards intelligent computational agent for professional architects. In: Proceedings of the IEEE/CVF conference on computer vision and pattern recognition, pp 13632–13641
15. Liu C, Wu J, Kohli P, Furukawa Y (2017) Raster-to-vector: Revisiting floorplan transformation. In: Proceedings of the IEEE international conference on computer vision, pp 2195–2203
16. Sigmund O, Maute K (2013) Topology optimization approaches: a comparative review. Struct Multidiscip Optim 48(6):1031–1055. https://doi.org/10.1007/s00158-013-0978-6
17. Burton H, Mieler M (2021) Machine learning applications in structural engineering: hope, hype or hindrance?. Struct Mag
18. Won J, Shin J (2021) Machine learning-based approach for seismic damage prediction method of building structures considering soil-structure interaction. Sustainability 13(8):4334

19. Schiftner A, Balzer J (2010) Statics-sensitive layout of planar quadrilateral meshes. In: Advances in architectural geometry 2010, pp 221–236. Springer, Vienna
20. Boden MA (1998) Creativity and artificial intelligence. Artif Intell 103(1–2):347–356. https://doi.org/10.1016/S0004-3702(98)00055-1
21. Chang AX, Funkhouser T, Guibas L et al (2015) ShapeNet: an information-rich 3D model repository. arXiv:1512.03012. https://doi.org/10.48550/arXiv.1512.03012
22. Perna V (2024) Emerging ecologies and the nascent paradigm: reconceptualizing architectural design in the current interregnum. The New that cannot be Born. In: Proceeding book of 4th international conference on scientific and academic research ICSAR 2024, pp 487–499. All Sciences Academy
23. Perna V (2024) Unveiling the post-digital paradigm cultural implications in a post-human design ecology. In: International conference 13–14 October 2023, Tirana-issues of housing, planning, and resilient development of the territory towards euro-mediterranean perspectives, pp 184–193. POLIS University Press
24. del Campo M (2022) Neural architecture: design and artificial intelligence. Applied Research and Design Publishing
25. del Campo M, Leach N (Eds) (2022) Machine hallucinations: architecture and artificial intelligence. John Wiley & Sons
26. Fritz J (2025) Understanding authorship in Artificial Intelligence-assisted works. J Intellect Prop Law Pract, 20(5):354–364.
27. Semjén ÁÁ, Szép J (2025) Integrating generative and parametric design with BIM: a literature review of challenges and research gaps in construction design. Appl Eng Sci. https://doi.org/10.1016/j.apples.2025.100253

Chapter 4
The LARGE(S): Defining the Urban Scale

Abstract The previous chapter investigated artificial intelligence (AI) in relation to its engagement with the architectural scale: the building as a designed artefact characterized by specific physical, cultural, and urban dimensions. The present one has the aim to address to the most complex spatial scale treated in this book: the city. The inner constitutive complexity lying in urban design problems makes them structurally different from the ones at the building design scale not just for their larger spatial extent: the city can be considered as a self-organizing system populated by millions of concurrent individual decisions that accumulated over decades and centuries, therefore every proposed design intervention in the city's palimpsest must deal with dynamics of emergence, path dependency, and social negotiation that refuse a reduction to simple optimization objectives. The present chapter explores the way machine learning models—convolutional neural networks (CNN), generative adversarial networks (GANs), graph neural networks (GNTs), and transformer-based architectures—are currently applied for the analysis, generation, and speculative reimagination of urban form. It focuses on four primary domains of application and research: morphological urban analysis and figure ground generation; urban street network analysis and mobility modelling; generative strategies for masterplans and urban layouts; and, finally, AI as a speculative and critical tool to reimagine the contemporary city. The discussion is grounded on three case studies extracted from the literature: Yap et al.'s Urbanity dataset and associated morphological analysis framework; Bettencourt's scaling law analysis of urban systems as a foundation for AI-based urban modelling; and del Campo, Manninger, and Carlson's Urban Fictions project, in which the combination between Nolli plan datasets and CNN-based style transfer on satellite images foresee new speculative urban morphologies that challenges the Albertian distinction between lineament and matter proposing alternative computation equivalents fluctuating between reality and utopia. The end of the chapter contains a synthesis of possible benefits, fundamental limitations, and the ethical and political dimensions of AI agency in the design of the city.

Keywords Urban AI · Urban morphology · Generative urban design · Convolutional neural networks · Style transfer · Urban fictions · GAN · Graph

P. Kyratsis et al., *AI Models for Design and Digital Prototyping*, Manufacturing and Surface Engineering, https://doi.org/10.1007/978-3-032-35144-9_4

neural networks · Master planning · Urban scaling laws · Smart City · Spatial analysis · Figure-ground · Nolli plan

4.1 Introduction

The scale hierarchy that was suggested in Chap. 1 and expanded upon in Chaps. 2 and 3 of this book gets its most intricate and significant representation in this chapter. First, it is important to understand that the city is not to be considered as a large building, nor much less as a simple collection and agglomeration of singe buildings but rather as a proper entity with its characteristics and overlapping dimensions. Therefore, if a building is the result of an intentional, directed, finite, and situated in time design process, the city is the result of a stratified historical process consisting in an accumulation of countless acts of construction, adaptation, and demolition on a timeline that goes beyond the one of any individual human being.

In one of his most famous formulations, Leon Battista Alberti compared the city to a large house, and the house to a small city: this affirmation is a convenient analogy in architectural theory for this chapter and also disguises the qualitative difference that matters most for the question of AI's applicability at the urban scale [1]. If a house can be completely designed, a city cannot. Nevertheless, this observation does not disqualify AI from a role in urban design. It redefines that role. In Chap. 3—indeed, at the building scale—AI-driven practices were investigated to understand its capacity to generate novel spatial configurations, predict structural performance, and implement productive formal estrangement in design practices. At the urban scale, some of the most significant contributions related to AI are analytical—it allows to read and interpret urban form, urban movement, and urban ecological performance at a resolution and scale that the human brain cannot match; generative—it allows the proposal of urban configurations not as an optimized solution but rather as hypotheses to establish a design dialogue; speculative—it helps to reimagine existing urban conditions through the application of machine vision to the large visual and cartographic archive behind the city. The next sections look at each of these contributions.

This chapter relies on an intellectual tradition that has the objective to identify the technical possibilities of AI at the urban scale without neglecting the humanistic complexity of the city. The urban map—to recall the opening of del Campo, Manninger, and Carlson's Urban Fictions essay [2]—is a cultural mainstay of the field of architecture: the abstract medium that most effectively conveys the intentionality of the urban project. From Alberti's lineament—the specific line in the mind of architect, intrinsically different from the matter of the building to Le Corbusier's firmness that architecture is a product of the mind, until arriving to Eisenman's difference between deep aspect and surface aspect, the planning process has always been depicted as a specifically and exclusive human cognitive act [2]. Therefore, the emergence of machine learning systems that are capable of generating full urban maps, interpolating between urban textures, and producing figure-ground outcome starting from datasets built on Nolli's map, raises also at this scale the same question on authorship and creativity that were addressed by del Campo at the building one through his Deep House project discussed in Chap. 3.

To untangle these conceptual knots, three case studies from the literature are woven into the discussion. Yap et al.'s Urbanity [3]—a global and open-source dataset of urban forms and functions that combines together street network morphology, land use, building footprints, point-of-interest data, etc.—proposes a data infrastructure for a broad set of opportunities for AI applications and serves to illustrate analytical dimension of AI at the urban scale. Bettencourt's scaling law analysis of urban systems [4] shows how a wide range of urban indicators, such as economic output, infrastructure length, crime rates, and innovation metrics, follows predictable power-law relationships with population size that present specific patterns across historical and cultural periods, and provides a theoretical foundation for AI modelling of urban systems that does not reduce the city to its designed components. Finally, Del Campo, Manninger, and Carlson's Urban Fictions [2] uses CNN-based style transfer to blend together images from SPAN's architectural renderings archive with satellite images of Vienna, Detroit, and other cities, and a large database of Nolli plans, with the aim of producing speculative urban morphologies that the authors ground on a philosophical discourse spanning from materialist cognition, the lineament tradition, and the posthuman design ecology.

The chapter is structured as follows. Section 4.2 defines the urban scale as a specific design domain, and examines the emergent, multi-stakeholder, and temporal characteristics that distinguish it from the building scale of Chap. 3. Section 4.3 presents a review of AI applications towards urban morphology analysis and figure-ground generation. Section 4.4 focuses on AI implementations for urban street network analysis and mobility. Section 4.5 explores generative approaches for the creation of urban layouts and masterplans. Section 4.6 studies AI as a critical and speculative design tool for cities through the references to the Urban Fictions project. Section 4.7 deals with ethical and political dimensions of AI agency in urban design. Section 4.8 synthesizes benefits, limitations, and open research directions. Section 4.9 provides a conclusion bridging to Chap. 5.

4.2 The Urban Scale: Characteristics and Design Challenges

To understand the difference between the architectural scale examined in Chap. 3 and the urban scale addressed here, the analysis has to be conducted on a qualitative perspective rather than on a merely quantitative one. This section explores the attributes of urban design problems that can impact both the potential contributions of AI to the field and also the fundamental limits behind its applicability.

4.2.1 *Emergence, Self-organization, and the Limits of Design*

One of the primary characteristics of urban form is that it is largely not design. In historic European cities, all the street networks, parcel configurations, building typologies, and land use patterns—the urban textures shown in del Campo's satellite images of Vienna, or in the famous eighteenth-century Nolli's map of Rome—are nothing other than the results of incremental, distributed, and widely uncoordinated acts of construction and adaption that stratified one on another century after century. The emergent spatial order behind these cities—in terms of their legibility, climatic performance, and pedestrian quality—cannot be intended as a direct intention of any architect but as the consequence of a contextual and historical building tradition that embedded structural and climatic intelligence in construction practices without formalizing it in design rules or books [5].

For the use of AI at the urban scale, this emergent character of urban form has two main implications. First, if AI models are trained with datasets built on the spatial patterns of historically formed cities, they can extract statistical regularities that rely on encoded design intelligence features—e.g., the relationship between building depth and height and the street width; the grain of a figure-ground pattern; and the density gradient that goes from the urban centre to the periphery—that never before were transcript as a proper explicit design knowledge. Second, the application of this learnt intelligence to the generation of new urban forms and configurations can reproduce the statistical side of historical urban forms but not the social, economic, and constructive condition that made their generations possible. Indeed, the speculative urban morphologies created by AI tools are nothing but a simulation of an urban character rather that a proposal for urban life, that is a fundamental distinction that urban planners and architects have to seriously consider when engaging with AI-generated urban proposals.

4.2.2 *Scale, Complexity, and Data*

Urban design relies on spatial scales—covering from a street block to a metropolitan region—that needs a volume and diversity of data that in no previous historical period could be gathered. Geographical Information Systems (GIS) datasets, archives from satellite images, mobile phone geo-localization features, energy consumption records, and air quality monitoring networks are unprecedented resources of information if properly applied to urban analysis and design. AI models—especially deep learning architectures that can analyse and extract meaningful set of data from an unorganized visual and spatial information—bear the intrinsic characteristic to exploit these precious resources, and can allow the generation of urban forms and analyses that were considered impossible a decade ago from a computational point of view [3].

Nevertheless, such richness in terms of urban data cannot be directly converted into design intelligence. In fact, urban data basically describes what currently exist, while urban design aims to propose what could and should exist. Therefore, the transcription from descriptive analyses of existing urban conditions and features to the proposal of new ones cannot happen without proper design judgement—a holistic understanding of what makes urban spaces liveable, ecologically sustainable, and culturally meaningful for their citizens—that cannot be extrapolated from data alone [6].

Currently, the main risk of uncritical AI urbanism approaches—namely, the direction translation of patterns derived from data into design proposals without a normative mediation—lies in the reproduction of already existing urban inequalities and the consolidation of historical injustices embedded into the spatial structure of the city. This criticality is examined more in depth in Sect. 4.7.

4.2.3 Time, Change, and the Urban Palimpsest

Cities have an ontological dimension that buildings do not. If a building is designed and built in a defined timeframe and stays unchanged for all its life, a city is an organism always in process and modification, simultaneously being built, adapted, inhabited, and demolished across its whole existence. As well-shown by Peter Eisenman in his architectural practice, the urban fabric is comparable to a palimpsest—a document being written over through time, in which earlier signs are readable across the new ones—and the intelligence needed to intervene in it is both historical and spatial [7]. An architect that designs a building inserted in an historical urban context must be aware of all the accumulated layers in the city's history, and understands the logics behind its different epoch, to propose an intervention that could establish a proper dialectic with the city's historical dimension.

AI models dedicated to urban analysis have started engaging the temporal dimension of urban form. For example, change detection algorithms are applied to time-series of satellite images to interpret the spatial dynamics of urban growth, densification, and are able to produce a temporal resolution and spatial coverage that previously was unthinkable [8]. Families of generative models trained on sequences of historical urban maps can understand the dynamics behind morphological changes, and allow the creation of plausible urban configurations addressing different development scenarios. However, the reading of urban history as a generative design resource—the understanding of the historical accumulated layers, and what they demand to the architect that intervenes in them—requires a rich contextual and interpretative intelligences that no current AI system yet shows.

4.3 Urban Morphology Analysis and Figure-Ground Generation

Urban morphology is the analytical discipline dedicated to studies related to the form of the cities, the street patterns, blocks, parcels, and buildings that compose them, and it is one of the most relevant conceptual tools for urban design practices and the domain where AI implementations have produced some of the most interesting and significant results. The figure-ground plan—grounded on Nolli's map tradition, where built mass is represented as a solid and open/public space as a void—is the main analytical instrument of urban morphological analysis, and the most directly accessible dimension for image-based AI methods.

4.3.1 Deep Learning for Urban Morphological Classification

Convolutional neural networks (CNNs) trained on urban morphology-based datasets have shown strong achievements on a wide range of urban classification and segmentation tasks spanning from the automatic cataloguing of urban morphological types from satellite images [9] to the segmentation of street network patterns (grid, radial, organic, etc.); from the identification of urban blocks understanding their spatial and programmatic character to the detection of informal settlements from aerial photography in cases where official documentation is missing. These classification capabilities are important especially for urban planning in developing areas, where a rapid and low-cost urban mapping is primary needed that conventional methods cannot offer at the required scale and speed.

Yap et al.'s Urbanity dataset [3] is one of the most significant contributions to the data infrastructure for urban morphological AI. The dataset collects morphological, functional, and accessibility data for cities worldwide—also using street network topology gathered from OpenStreetMap, land use classification from remote sensing, building footprint geometry from multiple open-source datasets, and point-of-interest density from commercial location data sources—and provides a multidimensional depiction of urban form and function that can be used both for comparative urban analysis and the training of AI models.

The authors have shown application of their dataset addressing the prediction of urban pedestrian accessibility from morphological indicators and the identification of urban typological clusters across cities of different cultures and developmental stages. The Urbanity dataset's global dimension and open availability make it a priceless resource for urban AI research, similar to what ShapeNet™ [10] represents for three-dimensional object AI at the design scale discussed in Chap. 2.

4.3.2 Generative Models for Urban Figure-Ground

Together with classification and analysis purposes, AI generative models are widely applied also to the synthesis of figure-ground configurations—that is the creation of possible footprints of building arrangements in a specific urban parcel or block. GAN models previously trained on dedicated figure-ground datasets from existing cities demonstrated the possibility to generate footprint configurations that are coherent with the spatial logic—in terms of density, grain, and open/public space structure—of a given urban typology, while also allowing urban designers to generate and evaluate multiple spatial and programmatic layout alternatives [11].

It is important to pay close attention to how these creative capacities relate to the heritage of urban figure-ground analysis in architectural practice and education. The figure-ground plan, starting from Nolli's Nuova Pianta di Roma (1748) to Colin Rowe and Fred Koetter's use of figure-ground analysis included in Collage City [7], has always been a primary instrument for reading the inner spatial logic of historic cities and generating design proposal that could be coherent to that logic. AI systems' capacity to apprehend spatial logics from Nolli plan dataset—indeed to generate figure-ground configuration that replicates the peculiar urban texture of Italian baroque urbanism, or the orthogonal grid structure of Haussman's Paris, or even the organic grain or a Medieval city—poses meaningful question of what is the real meaning of understanding urban spatial logic, and also if statistical pattern learning from existing plan data can be considered as a proper understanding of it or not.

Such a question is exactly one of those that del Campo, Manninger, and Carlson aim to address with their Urban Fictions project [2], from a perspective that is grounded as much in architectural theory as a in machine learning one. Their use of CNN-based style transfer applied to databases of Nolli plans—and producing figure-ground compositions that are recognizable in their spatial texture even though globally estranged because of embedding urban traditions that no single city has—is to be intended as a design experiment that addresses the limits of this question through the visual representation of its implications. The produced images, in which the features of historic figure-urban urbanism appear in a form that could not have accumulated organically through time, are a kind of proof by demonstrations that AI has learnt and incorporated something concerning urban spatial form that is genuine, transferrable, and reproducible, even though not properly mastering about the social, material, and historical processes through which that form was produced.

4.4 Urban Street Networks, Mobility, and Graph-Based AI

A street network is one of the main spatial infrastructures of a city: the system of paths, roads, and intersections that organizes movements, defines the public realm, and constructs the relationships between built systems and open/public space. Also in this case, urban street networks exhibit different characteristics in terms of spatial and topological structure related to cultural and historical differences—from the ancient Greek's Hippodamian system to the Roman cardo-decumanic one; from the organic patterns of the Islamic medinas to the radial scheme of the Baroque city—that resonate in different models of public life, different theories of urban order and organization, and different responses to climatic and topographic conditions.

4.4.1 Graph Neural Networks for Street Network Analysis

The natural representation concerning street networks is a graph where nodes correspond to intersections, edges to street segments, and the network's spatial features are embedded in the topological and geometric properties of the graph.

Graph neural networks (GNNs)—specific deep learning architectures created to operate on graph-structured data that can learn representation that can identify both the node-level characteristics and the structural relationships between the single nodes—have been applied to the analysis of urban street networks showing results that significantly improve the field of quantitative urban morphology [12].

Boeing's OSMnx frame [13] created a methodological infrastructure for a large-scale acquisition, analysis, and comparison of urban street networks extracted from OpenStreetMap that allowed a systematic comparative study of street network morphology among hundreds of different cities. When applied to this data infrastructure, GNN family models have exhibited the ability to predict urban mobility patterns from street network topology—for example, they estimated the traffic flows (both pedestrian and vehicular) from the network itself, without the need for direct measurement—and to identify specific network configurations that could be associated with positive urban results such as mixed land use intensity, walkability, and level of social interaction density [14]. It is clear how these predictive capacities have a direct implication for urban design conventional practice: they allow designers to evaluate the mobility consequences of a network configuration that they propose without spending time and money in crafting simulations models of comparable complexity.

4.4.2 Urban Scaling Laws and AI Urban Modelling

The urban scaling law research, initiated by Bettencourt et al. [4] and developed in Bettencourt's The Origins of Scaling in Cities [4], represents a theoretical foundation for the AI-based modelling of urban systems that goes beyond the analysis of individual street networks or morphological datasets. This work argues that a wide range of urban indicators—economic productivity, innovation rates, infrastructure length, energy consumption, crime rates—compose a consistent power laws of city population that can be found across cultures, historical periods, and levels of development.

For instance, urban economic growth and innovative outputs scale super linearly with population (with exponents of approximately 1.15), while infrastructure quantities scale sublinearly (with exponents of approximately 0.85), implying that larger cities are simultaneously more economically productive per capita and more infrastructure-efficient.

The underlying social and spatial dynamics of urban agglomeration—the rise in interaction rates and, consequently, in the rate of economic and cultural exchange that takes place as urban density increases—are reflected in these scaling connections, which go beyond simple statistical regularities. In AI-based urban modelling, such scaling laws provide a system of quantitative constraints that each generative urban model must satisfy if the outputs aim to depict plausible cities rather than arbitrary sets of spatial configurations. A GAN trained to generate urban layouts that can consider the scaling relations between the density of the population, the length of the street network, and the intensity of economic activity, might be able to produce outputs with a level of systemic-based urban realism that no pure and exclusive morphological training can achieve [15].

For this reason, an active field of research is the integration of scaling law constraints in urban AI generative models. Such approaches deal with the use of scaling law-derived performance metrics as objectives or constraints within evolutionary optimization frameworks for urban layout; the creation of multiscale generative models that preserve scaling consistency across the range of spatial scales from the street block to the metropolitan region; and the training of physics-informed neural networks that incorporate scaling relationships as regularization terms in the model training objective [16]. These approaches are the beginning of a more informed AI-based urbanism where the systemic character of a city is respected, and where the generated outcomes understand urban form not as a mere spatial pattern but also as a complex system with dynamic implications and interactions.

4.5 Generative Approaches to Urban Layout and Master Planning

The most direct engagement of AI with professional practices related to urban design is the application of AI generative methods towards urban layout and master planning: the new design of new city portions or urban quarters; the restructuring of existing ones; and the planning of brand-new settlements. This domain is also the one where the distance between the technical capability of AI and the professional requirements is currently larger, and where the ethic and political implications of AI agency are highest.

4.5.1 From Building-Scale Generation to Urban-Scale Layout

Chaillou's ArchiGAN [17] and Nauata et al.'s House-GAN [18] are some of the existing floor plan generative systems that work at the scale of the individual apartment or generic dwelling. To scale up these processes at the urban scale means not only focusing on the creation of individual room configurations but of parcel organizations, block structures, street network layouts, and the relationship between built mass, public space, and programmatic allocation along a specific urban area. The amount of design variables, the variety of constraint types, and the interdependence between decisions at various spatial scales all significantly rise in relation to the building size as a result of this extension, which also brings a qualitative increase in problem complexity.

Graph-conditioned generative models have been proposed as an option for urban-scale layout generation. They extend the House-GAN approach from a graph based on room adjacency to an urban program graph where the nodes represent specific urban zones (residential, commercial, civic, green) and the edges symbolize the relationship between them. This family of models learn to produce spatial configurations that can satisfy designated zoning relationships from training data that are extracted from existing master plans, with the intent of enabling urban designers to be able to specific programmatic intensions in their project as relational graphs so that they can receive an informed candidate spatial configuration to evaluate and then further develop. Current practical limitation in existing systems is their restriction to only elaborate simple program graphs and regular site geometries. Therefore, it is still a research problem to extend urban design practice to the complicated, irregular, and topographically variable locations.

4.5.2 Multi-Objective Urban Layout Optimization

Multi-objective evolutionary algorithms, in combination with neural network surrogate models for urban performance evaluation, have been used for the optimization of urban block layouts to predict simultaneous performances concerning natural ventilation, daylight access, walkability, and ecological connectivity objectives.

Researchers such as Ferroni et al. [19] showed that urban block configurations that significantly outperform traditional planning practice on environmental performance metrics while maintaining program compatibility and buildability requirements can be found using a genetic algorithm in conjunction with surrogate models for solar radiation and natural ventilation. This method was expanded by Larkin et al. [20] to simultaneously optimize block compactness, pedestrian catchment area, and street network connection. They showed Pareto-optimal trade-offs between these goals in a variety of urban density situations.

To properly integrate this optimization tools into consolidated urban practices, there is the need to develop workflow interfaces that can make multi-objective urban optimization understandable and manageable from urban planners and architects without the need for special computational expertise: a requirement related to the interface challenge in regard to product design even though, in this case, characterized by a greater complexity and a longer time horizon of urban design decision. Platforms such as CityEngine, parametric design environments like Rhinoceros with Grasshopper, and novel urban design tools based on geodesign frameworks have started incorporating AI-assisted performance evaluation, but the integration of AI generative capabilities in these environments still remains at a prototype phase.

4.5.3 Procedural Generation and Urban Growth Simulation

Procedural generation methods are rule-based systems able to generate spatial configurations through the recursive application of production rules. They have a long history concerning urban simulation: from the Cellular Automata models of urban growth to agent-based models dealing with land use modifications. These procedural frameworks' use of machine learning has resulted in hybrid systems that combine the data-driven adaptability of learnt models with the interpretability of rule-based generation. CityEngine [21], developed by Parish and Müller, is one of the most widely known and used urban procedural generation systems that uses a shape grammar-based approach able to generate three-dimensional city models from given zoning and street network data. This example shows how the integration of neural networks trained on urban morphology datasets—through a shape grammar system—can lead to the automatic extrapolation of rules from existing urban patterns without manual specification.

Furthermore, deep learning-based urban growth simulation models—trained on a time-series of satellite images to predict possible dynamics for urban growth—have demonstrated the ability to generate accurate planning scenarios especially for what concerns rapidly urbanizing contexts [8]. Such systems learn the spatial dynamics of urban growth from patterns of historical expansions and produce plausible options for possible futures considering different population growth, infrastructure investment, and land use regulation scenarios. Considering this, their applications represent a significant AI contribution to the field of urban design especially in those contexts where formal urban planning capacity is not enough developed and urban growth is rapid such as sub-Saharan African peri-urban areas, Southeast Asian coastal cities, and South American informal settlement zones.

4.6 AI as Speculative Instrument: Urban Fictions and the Reimagination of the City

The applications of AI presented in the precedent sections addressed primarily an analytical or optimizing dimension: they basically read existing urban patterns and then predict performance outcomes while generating specific configurations to satisfy some given constraints. This section has the aim to present a qualitatively different modality through which AI can be linked to the city, highlighting its use as a speculative and critical instrument to reimagine urban form. In this perspective, AI is engaged to produce urban images that do not want to propose solutions for defined design problems but rather to open new questions concerning the ontological dimension of urban cognition; the character of the city; and the possible relation between human and machine intelligence for urban design purposes.

4.6.1 The Lineament and the Algorithm: Del Campo's Urban Fictions

The Urban Fictions project [2], developed by Matias del Campo, Sara Manninger, and Alexandra Carlson, is presented by its writers with a question that serves as both a theoretical assertion and a provocation: what is the position taken by architectural culture when the use of machine learning processes gives birth to results that might challenge the exclusive position of human intelligence in terms of authorship? The question is posed in regards to the background in architectural tradition that follows the idea of the lineament—from Alberti's distinction, in De re aedificatoria, between the line present in the architect's mind and the material of the building, to Le Corbusier's persistence on the idea that architecture is a product of the (human) mind, to Eisenman's separation between deep aspect and surface aspect [2]. Therefore, following this tradition, it is clear how planning process can be intended as a specific cognitive act: the expression of human intellect and intentionality in a spatial form.

Urban Fictions uses an urban design experiment to examine this heritage rather than just question it. From a technical perspective, the project uses neural style transfer—a CNN technique where a convolutionary network extracts the stylistic features of a reference image to then recombine them on the structure of a target image [22]—to two different sources for urban images. In the first set of experiments, the content photos are SPAN's own architectural renderings, while the style images are satellite images of already-existing cities, such as Detroit and Vienna; the CNN's characteristic mediates the combination and produces spatial topologies from real urban environments that are structured on the formal language of SPAN's speculative architecture.

In the second series, Nolli's plans are used in a large database from the training material, and the CNN produces figure-ground configurations that mix the characteristic textures of several historical urban traditions and generate outcomes that could have not been accumulated and stratified organically through time.

What is important to understand in this case is that all the results are not design proposals in any conventional sense, and, as described by the authors, they constantly oscillate between tangible realism and unlikely scenarios thus between reality and utopia. Furthermore, they do not specify any street, building, or program but they rather are urban fictions: hypothetical depictions of nonexistent cities created by a machine learning method that has discovered something true about the spatial nature of historic urbanism without comprehending the social, material, and historical processes that gave rise to that character. In their work, the authors rely on a materialistic philosophical frame—basically connecting to the tradition that sees the act of thinking as a material process in the brain, a set of neurochemical reactions able of forming concepts—to highlight how the difference between human and machine cognition is just a substrate distinction: if the first uses neurochemical processes to produce thought, the second one is based on computational ones.

The produce urban fictions have some qualities that connects to the previous del Campo's speculations—through the Deep House project discussed in Chap. 3—on estrangement that he defines as 'wonderfully strange': since the computer perceives urban form without the cultural presumptions and typological standards that shape human perception of the city, it generates morphologies, chromatics, and spatial organizations. The image of Vienna reframed through SPAN's design language is not Vienna but, at the same time, it is not even SPAN's architecture; it is a third thing—a Calvinoesque city—deriving from the combination of two inputs of spatial intelligence, one human and the other computation, through the use of a neural network that grasps their shared formal vocabulary but do not consider their separate cultural contexts. The spatial trauma of a city in industrial decline is blended with the formal ambition of speculative architecture in ways that hover between critique and utopia in the image of Detroit processed using the same method. This urban image makes visible, by its very strangeness, the difference between the city that exists and the city that could be imagined.

4.6.2 The Nolli Plan as Training Data: Urban Memory and Machine Vision

From an architectural-historical perspective, the use of Nolli plan structured databases as a training set for CNN-based urban figure-ground generation in Urban Fictions is one of the most theoretically and rich implications aspect of the project.

Giambattista Nolli's Nuova Pianta di Roma, published in 1748, is one of the most famous and celebrated cartographic instruments in Western urbanism: basically, it is a figure-ground map that, with extreme precision, distinguishes the solid mass of Rome's built environment from the voids of its streets, piazzas, and publicly accessible interiors such as churches, colonnades, and institutional buildings, that for Nolli are an explicit component of the public realm of the city. Nolli map gave birth to a tradition of figure-ground urban analyses that have influenced many urban design theories such as Giurgola and Mehta's contextual urbanism or Colin Rowe and Fred Koetter's Collage [17], until arriving at the contemporary emphasis posed by contemporary New Urbanist on the fundamental role of public space.

It is clear then how when del Campo, Manninger, and Carlson train a CNN on a Nolli plan database for generative purposes; they are basically exploring that the machine learns from a consolidated tradition of urban representations.

The answer, contained in the produced outcomes, is basically what a human urban design could extrapolate from Nolli's work but with a machine perceptual capacity that exceeds human possibilities and that can generate hundreds of plans simultaneously, and identify spatial patterns at different scales and densities. Nevertheless, since the machine learns from Nolli map the spatial form without understanding its social content—the distinction between public interior and private enclosure that gives the Nolli map its analytical significance—the resulting figure-ground analyses are depleted of this layer that gives to the original map its peculiarity of being a tool for thinking about urban publicness rather than just a urban geometry representation.

Therefore, the figure-ground alternatives generated by the CNN are formally plausible as urban plans but they are not intrinsically valid proposal intersecting the public realm.

This specific gap—between the formal learning of the machine and the social intelligence of the urban designer—is considered by the authors as a productive feature rather than a disqualifying one. The machine-generated outcomes serve as an intellectual stimulus and not as a completed urban proposal: they are urban images that provoke questions about possible cities; what spatial structures architectural archives encode, and what the possible findings of a design intelligence are (the machine one) that operates without typological conventions. In this way, Urban Fictions is an experiment in both architectural epistemology and AI application, exploring the question of what urban knowledge is and what sorts of machines can help expand it.

4.6.3 Speculative AI Urbanism and the Discipline of Architecture

The speculative dimension behind Urban Fictions is situated in a current field of research in architectural practice where AI is considered not a mere optimization tool but as a generative instrument to expand the horizon of possibilities of the discipline. Similar experiences can be found in Refik Anadol's work, whose data-driven architectural installations attempt to materialize the hidden pattern features behind urban datasets in the form of a spatial experience; in the research of Neil

Leach on machine hallucinations and the connection between AI and architectural imagination [23]; and other AI-based research that do not consider anymore computer as a simple calculator but as a proper active counterpart that can contribute through intelligence in the design process.

The main difference between the most intellectually rich examples of speculative AI urbanism—such as Urban Fictions—from the burgeoning number of AI-generated architectural images is the existence of an intellectual framework that tries to situate those images in a broader discourse addressing the nature of architectural intelligence, the limits of computational cognition, and the conditions when machine-generated form can be used as a proper design stimulus from architects and urban designers. Therefore, the outcomes in Urban Fictions are not a graphic demonstration of AI capacity but rather a tool for architectural thought. They present an argument that is both architectural and computational in nature. They discuss the connection between the lineament and the algorithm, the urban archive as a source of machine-readable design intelligence, and the beneficial estrangement that machine vision brings to architectural culture.

4.7 Ethical and Political Dimensions of AI Agency in Urban Design

The ethical and political dimensions of AI urban design are constitutive of the technical questions themselves, and should be addressed after the technical questions behind these procedures have been resolved. Indeed, the communities whose interests are represented in the training data, the data used to train urban AI models, the goals those models are intended to optimize, and the institutional frameworks within which AI-generated urban proposals are assessed and put into practice are all value-laden decisions that impact the social outcomes of AI urbanism just as much as any technical design choice.

4.7.1 Representation, Bias, and the Training Data Problem

As previously described in this chapter, current tools for AI-assisted urban morphology analysis and generation are trained on existing datasets related to urban form—among others, satellite images, street network data, figure-ground plans, and building footprint databases. All these datasets are the results of historical urban processes that were shaped under specific conditions of social power, economic inequality, and also cultural exclusion that are maybe not clearly understandable from data artefacts describing urban space but do exist in their spatial consequences. For example, a CNN trained on figure-ground plans of European and North American city can learn the spatial logic behind these cities but not the historical process of segregation, displacement, and inequality that conformed them; therefore, the model can exhibit the tendency to reproduce the spatial disparities lying in the training corpus through which the design proposals are generated.

In this direction, even though the representational bias in urban AI training data is a consistent part of AI literature in urban planning [27], an adequate solution still remains blurry and unclear. One of the most proposed approaches is based on diversifying the training datasets so that they also embed urban forms from wider cultural, geographic, and political context. Nevertheless, this solution addresses the symptoms without addressing the main problem: that the social intelligence needed to produce urban plans that are inclusive, equitable, and sensitive to the unique circumstances of a given community cannot be encoded by any dataset of built urban form, no matter how diverse.

Indeed, for how comprehensive data diversity can be, the addition that societal participation brings to the design process—local knowledge, contextual knowledge, and political accountability—is still not replaceable by any existing AI system.

4.7.2 Participation, Agency, and the Democratic City

The possibility to include participatory in AI-assisted urban design—processes where the community is actively involved in the definition of design objectives, the evaluated of the produced alternatives, and the selection of a set of preferred solutions—is an important direction in this field in terms of both technical and political dimensions. From a technical perspective, participatory AI urbanism requires the creation of accessible interfaces through which community members can enter in contact and engage with AI-generated urban proposals: platforms that can translate spatial configurations and properties into experiential representation (street-level visualizations, walkability scores, noise environment simulations) that support citizens in evaluating the design solution without the need for a specific architecture or urban design literacy [24].

In terms of politics, participatory AI urbanism necessitates institutional frameworks where community involvement is decisional rather than merely consultative; that is, where the community's assessments influence the optimization goals of AI systems instead of presenting AI-optimized solutions to communities for approval. The development of these frameworks is both a political and institutional challenge that needs the coordination of multiple stakeholders among AI researchers, urban designers, architects, planning practitioners, and local communities.

On a research perspective, the technical dimension of this issue—find a proper way to embed community preferences as AI optimization objectives—is an active field of research that spans from preference modelling to multi-objective optimization.

4.7.3 AI and the Right to the City

The concept of the right to the city—introduced by Henri Lefevbre and addressing the right of urban inhabitants not only to use the space they live but also to participate in the production of the space itself and shape the city around their collective desires and needs [25]—is a normative framework that can serve to address the ethical implications of AI implementation in urban design processes. Indeed, if AI urbanism produce functional outcomes that do not into take the needs of the communities that will inhabit them, that is, in Lefevbre's own words, a denial to the right to the city since it treats citizens as one of the objects of the design process rather its subjects, and only tends to reproduce—in a computational disguise—a top-down planning tradition that produce some of worst urban interventions of the last century.

To avoid this, it is fundamental that AI urbanism is used as a tool that can amplify community design agency and not replace it and interpret the complexity or urban systems to support community agency about urban futures. The idea is to shape a form of AI urbanism that could be coherent with the inner humanistic values of the discipline of architecture and that have been part of its foundation. In this regard, the speculative images generated in the Urban Fictions project, situated between utopia and reality, between the city as it is and as it could be, is a possible direction that points towards a more open, critical, and humanistic way to use AI to engage with the urban: a modality that uses machines' capacity for pattern recognition and formal combination not as a simple instrument to optimize the city as it is, but to create a design framework to explore an imaginative territory where collective urban futures can be designed, and collectively debated and evaluated.

4.8 Benefits, Limitations, and Open Research Directions

4.8.1 Demonstrated Benefits

There are several categories of benefit that can be identified in the application of AI families at the urban scale. In urban morphological analysis, deep learning models have allowed the generation of automated classifications and urban segmentations from satellite images covering a territorial and temporal range that was previously impossible for human intellect; furthermore, they have supported urban monitoring, the mapping of informal settlements, and comparative urban morphology research at a speed space that traditional survey methodologies could not allow [3, 9]. Neural network surrogate models and multi-objective evolutionary algorithms have found urban layout configurations that outperform traditional planning practice on environmental performance metrics, such as solar access, natural ventilation, and walkability, in urban performance modelling [19, 20]. This shows that AI-assisted urban layout optimization can contribute quantifiable performance improvements relative to current practice.

Regarding urban mobility analysis, GNN-based models trained on street network data have produced accurate prediction of pedestrian and vehicular flow distributions starting from network topology alone, and allowed urban designers to assess and evaluate mobility consequences of specific network configurations without the need to build complicated and expensive simulation models [14]. In speculative design, the Urban Fictions research showed that CNN-based style transfer can be applied to urban images and cartographic archives to produce generative urban images of aesthetic and intellectual interest that can concretely contribute to the on-going architectural discourse concerning the city, machine cognition, and the limits of conventional lineament tradition in urban design.

4.8.2 Principal Limitations

The current limitation of existing AI systems applied at the urban scale are undoubtedly more significant than the ones already discussed in Chap. 3 concerning the architectural scale, and they reflect the intrinsic complexity of the city as a design object. One of the most significant drawbacks is the gap between the statistical and morphological intelligence that AI models can acquire from urban-based datasets and socio-political and historical intelligence needed to design the city, that is the real understanding of what an urban space means for its inhabitants, the conditions of inclusiveness and justice it should embed, and what kind of relationship between the private and the public sphere it should produce. Furthermore, this existing gap cannot be considered a mere inadequacy of current AI systems that could be solved with more training or improving the model capacity, but it is the reification of a categorical distance between statistical patterns and contextual design intelligence [6, 26].

The applicability of AI-generated urban proposals is limited in situations where design innovation is most urgently needed: communities that have experienced historical disinvestment, displacement, and spatial exclusion. This is due to the data representativeness limitation, which states that urban AI training datasets reflect the historical built outcomes of unequal social processes.

The temporal limitation—that is the difficulty of training AI models on urban processes that are stratified over decades and centuries in comparison to the shorter timeframes synthetized in available datasets—confines the applicability of AI-based models for urban growth in relation to the long-term horizons that urban design decision requires. Finally, the participation limitation—the lack of meaningful community engagement to specify the AI generative constraints and training objectives—poses serious limits to the democracy behind AI-generated urban proposals, despite of their technical rigour.

4.8.3 *Open Research Directions*

Three are the interesting research directions for AI at the urban scale that are worthy of mentioning. First, and most important, is the development of a socially and historically contextualized urban AI that could reflect the social dimension of the urban fabric and not only its geometric one. For fulfilling this objective, it is fundamental for the cooperation among machine learning experts and urban sociologists, historians, and political economists to realize proper representations of urban spaces that can accommodate social meaning together with spatial structure, and to develop training methods that can proficiently learn from both quantitative and qualitative sets of urban data [27].

Second is the investigation of the possibilities offered by participatory urbanism while making it technically accessible through workflows in which the community members shape the objectives and evaluate AI outcomes at the urban-scale generation. To achieve that, there is the urgency to produce advancements in the field of explainable AI (that is the capacity for AI systems to be understood and readable also from non-specialist users), in interface design (the adaptation of spatial AI results into formats and representations that could be readable from the community), and in institutional design, through the creation of planning frameworks where AI-assisted community design could have its own political weight.

Third, and final, is the development of urban AI that can be temporally aware: systems that investigate urban changes considering time horizons (the centennial cycle of urban morphological evolution, the daily cycle of use, and the decadal cycle of construction and adaptation) that are at the core of the temporal structure of the city. To do so, there is the need for novel approaches to represent and learn about temporal dynamics through the combination of historical cartographic data, satellite time-series imagery, and contemporary sensor networks.

4.9 Conclusions

The current use of machine learning models for the analysis, creation, and speculative reimagining of the city has been examined in this chapter. These models include convolutional neural networks, graph neural networks, generative adversarial networks, multi-objective evolutionary algorithms, and reinforcement learning systems.

The urban scale is considered the most consequential and complex of the domains addressed in this volume: a specific domain where the design and the undesigned cannot be separated; where spatial form is the collective production of social procedures rather than just design intentions; and in which AI's most important contributions are still at an analytical and speculative level rather than a generative one in a professional dimension.

The discussion has been structured around three case studies drawn from the existing literature. Yap et al.'s Urbanity [3] served as an example of a framework for understanding how global and multidimensional datasets can allow AI-driven morphological analysis under the light of new scope and resolution. Bettencourt's urban scaling laws [4] explained the need for an AI urban modelling that can interpret the systematic character of the city—in terms of emergent economic, social, and spatial dynamics—instead of understanding urban form as a mere spatial pattern detached from the urban life component. The most philosophically rich of the three, Del Campo, Manninger, and Carlson's Urban Fictions [2], situated CNN-based style transfer applied to satellite imagery and Nolli plan databases within the architectural tradition of the lineament interrogates the question of what a machine can know about the city and how that knowledge contributes to the architect's ability to imagine urban futures through the medium of speculative urban images that oscillate between utopia and reality.

Furthermore, the political and ethical dimension of AI-based urbanism has been analysed in comparison to its technical possibilities, highlighting the conviction that the implementation of AI agency in urban design cannot be separated from the topic of democratic agency in and for the city. Therefore, a politically uncritical AI urbanism—used to optimize the geometric features of urban form without engaging the social processes that produce it and that will accept or not the AI-based designed interventions—will only repeat the well-known failures of twentieth-century technocratic planning only under a new computational perspective. The alternative, briefly presented at the end of this chapter, is an AI urbanism that embeds and amplifies human and community design intelligence and that uses machines' analytical and generative capabilities under the aegis of more democratic, equitable, and participatory urban futures.

At this point, Chap. 5 moves away from the design of urban form and points towards the platforms and ecosystems through which AI design practices are developed, distributed, and integrated into professional practices, and looks at the institutional arrangements, software infrastructures, and API frameworks that decide how designers and planners operating at each of the spatial scales covered in this volume can access the capabilities discussed in Chaps. 2, 3, and 4.

References

1. Alberti LB (1988) On the art of building in ten books. Translated by Rykwert J, Leach N, Tavernor R. MIT Press, Cambridge MA
2. del Campo M, Manninger S, Carlson A (2022) Urban fictions. Machine learning and the city: applications in architecture and urban design, 183–188. https://doi.org/10.1002/9781119815075.ch14
3. Yap W, Stouffs R, Biljecki F (2023) Urbanity: automated modelling and analysis of multidimensional networks in cities. npj Urban Sustain 3(1):45
4. Bettencourt LMA (2013) The origins of scaling in cities. Science 340(6139):1438–1441. https://doi.org/10.1126/science.1235823

5. Hillier B, Hanson J (1989) The social logic of space. Cambridge University Press
6. Jacobs J (1961) The death and life of great American cities. Random House, New York
7. Rowe C, Koetter F (1978) Collage City. MIT Press, Cambridge MA
8. Liu X, Hu G, Chen Y et al (2018) High-resolution multi-temporal mapping of global urban land using Landsat images based on the Google Earth Engine platform. Remote Sens Environ 209:227–239. https://doi.org/10.1016/j.rse.2018.02.055
9. Srivastava S, Vargas-Muñoz JE, Tuia D (2019) Understanding urban landuse from the above and ground perspectives: a deep learning, multimodal solution. Remote Sens Environ 228:129–143. https://doi.org/10.1016/j.rse.2019.04.014
10. Chang AX, Funkhouser T, Guibas L et al (2015) ShapeNet: an information-rich 3D model repository. arXiv:1512.03012
11. Huang W, Zheng H (2018) Architectural drawings recognition and generation through machine learning. In: Proceedings of the 38th annual conference of the association for computer aided design in architecture, Mexico City, Mexico, pp 18–20
12. Scarselli F, Gori M, Tsoi AC, Hagenbuchner M, Monfardini G (2009) The graph neural network model. IEEE Trans Neural Netw 20(1):61–80. https://doi.org/10.1109/TNN.2008.2005605
13. Boeing G (2017) OSMnx: new methods for acquiring, constructing, analyzing, and visualizing complex street networks. Comput Environ Urban Syst 65:126–139. https://doi.org/10.1016/j.compenvurbsys.2017.05.004
14. Khairy M, Mokhtar HM, Abdalla M (2025) Adaptive traffic prediction model using Graph Neural Networks optimized by reinforcement learning. Int J Cognit Comput Eng 6:431–440
15. Bettencourt LMA, Lobo J, Helbing D, Kuhnert C, West GB (2007) Growth, innovation, scaling, and the pace of life in cities. Proc Natl Acad Sci USA 104(17):7301–7306. https://doi.org/10.1073/pnas.0610172104
16. Batty M (2013) The new science of cities. MIT Press, Cambridge MA
17. Chaillou S (2020) Archigan: artificial intelligence x architecture. In: Architectural intelligence: selected papers from the 1st international conference on computational design and robotic fabrication (CDRF 2019), pp 117–127. Singapore: Springer Nature Singapore
18. Nauata N, Chang K-H, Cheng C-Y, Mori G, Furukawa Y (2020) House-GAN: relational generative adversarial networks for graph-constrained house layout generation. In: Vedaldi A et al (eds) Computer vision—ECCV 2020. Springer, Cham, pp 162–177. https://doi.org/10.1007/978-3-030-58452-8_10
19. Ferroni S, Ferrando M, Carlucci S, Causone F, A combinatorial multi-objective optimisation model to support the design of urban renewable energy communities. Available at SSRN 5370306
20. Larkin A, Gu X, Chen L, Hystad P (2021) Predicting perceptions of the built environment using GIS, satellite and street view image approaches. Landsc Urban Plan 216:104257
21. Parish YI, Müller P (2001) Procedural modeling of cities. In: Proceedings of the 28th annual conference on Computer graphics and interactive techniques, pp 301–308. https://doi.org/10.1145/383259.383292
22. Gatys LA, Ecker AS, Bethge M (2016) Image style transfer using convolutional neural networks. In: Proceedings of the IEEE conference on computer vision and pattern recognition (CVPR), Las Vegas, pp 2414–2423. https://doi.org/10.1109/CVPR.2016.265
23. del Campo M, Leach N (Eds) (2022) Machine hallucinations: architecture and artificial intelligence. John Wiley & Sons
24. Salama AM (2016) Spatial design education: new directions for pedagogy in architecture and beyond. Routledge
25. Lefebvre H, Kofman E, Lebas E (1996) Writings on cities. Mass, Cambridge
26. Kitchin R (2014) The real-time city? Big data and smart urbanism. GeoJournal 79(1):1–14. https://doi.org/10.1007/s10708-013-9516-8
27. Sanchez TW, Brenman M, Ye X (2024) The ethical concerns of artificial intelligence in urban planning. J Am Plan Assoc, 91(2). https://doi.org/10.1080/01944363.2024.2355305

Chapter 5
Platforms and Ecosystems for AI-Driven Designing and Prototyping

Abstract AI-driven design and prototyping at an increasing spatial scale—from the designed product to the building to the city—were covered in the previous chapters of this collection. In each of the presented scale, a repeating structural pattern was identified: the current capabilities of AI-based tools are not standalone systems but they all word as a component in broader software ecosystems. Platforms, APIs, scripting environment, fabrication pipelines, etc. create a mediation between the designer/architect's workflow and the model's analytical or generative capacities. This chapter aims to examine such platforms and ecosystems, and investigate which are the existing computational infrastructures through which AI design peculiarities are developed, integrated, and applied in professional practice. The design-to-fabrication technology stack is broken down into three branches: visual programming platforms and parametric design environments, where AI plays a consistent role in familiar design tools using plugins and API integration; AI-native creative platform that are based on designer-oriented and familiar interfaces to give direct access to the model families capabilities; and, last, fabrication-integrated AI pipelines that have the objective to directly connect the model's outputs to digital manufacturing systems both relying on additive and subtractive processes. Three examples from the literature show how these platform levels function in real-world scenarios: Oxman's Material-Based Design Computation framework, which created a fundamental framework for incorporating material intelligence into platforms for computational design; the well-documented in literature Rhinoceros™-Grasshopper™ ecosystem, which is one of the most favourite platforms for research-level AI integration dedicated to the fields of architectural and product design; and the CAD-based programming study by Kyratsis et al., which allows AI-driven process automation to be methodically incorporated into industrial CAD/CAM systems for manufacturing applications. At the end of the chapter, benefits, limitations, and some critical questions concerning current platform design agency and dependencies are synthesized.

Keywords Design platforms · CAD/CAM ecosystems · Grasshopper™ · Parametric design · API integration · AI-native tools · Fabrication pipelines · Additive manufacturing · Generative design · Computational design · Design automation · Digital fabrication

P. Kyratsis et al., *AI Models for Design and Digital Prototyping*, Manufacturing and Surface Engineering, https://doi.org/10.1007/978-3-032-35144-9_5

5.1 Introduction

In each chapter of this volume, a specific design scale has been used to describe the current panorama of AI-assisted practices in design and prototyping, and the most diffused AI models, workflows, and design challenges. The platform infrastructure that enables the access and deployment of AI capabilities is what these descriptions have consistently assumed without making it the main focus of analysis: scripting APIs, cloud computation services, generative creative tools, parametric design environments, and digital fabrication systems. All of these are part of the technological substrate that composed contemporary design practices, and this very infrastructure is the primary subject of this chapter whose importance cannot be underestimated in regard to real-world use of AI in design.

Indeed, all the AI capabilities that are directly accessible or integrated in the software that designers use are more research curiosities rather than practice realities. The capabilities most likely to influence the nature of AI-assisted design across design disciplines in the near future are those that are integrated into well-known design environments, respect professional practice conventions, and lower the barrier to engagement without diminishing the depth of what is possible. Therefore, it is more than needed to understand the existing platform landscape—its current state, development directions, and the limitations, benefits, and dependencies it introduces—to assess the current agency of AI in the design field.

The design-to-fabrication technology stack, a conceptual framework that organizes the analysis across the several platform categories pertinent to design and prototyping, is the basis for the chapter's three tiers.

The first investigates visual programming platforms and parametric design environments such as Rhinoceros™ with Grasshopper™, Autodesk™ Dynamo™, Bentley™ GenerativeComponents, and their equivalents in industrial CAD platforms, where AI tools are integrated within familiar design environments and alongside traditional geometric operation tools. The second delves into AI-native creative platforms, considering the increasing ecosystem of generative AI tools—image diffusion platforms, three-dimensional shape generation services, large language model APIs—that gives access to AI capacities through the creation of interface dedicated to creative professionals rather than engineers. The third, and final, explores the possible integration of AI pipelines in digital fabrication and manufacturing and focuses on systems that can guarantee the direct connection of AI-designed geometries to physical manufacturing processes such as CNC machining, additive manufacturing, robotic fabrication.

To ground the discussion in practice, three are the case studies drawn from existing literature. A foundational model for the integration of material intelligence—the knowledge of how material properties shape design possibilities—within computational design platforms based on the work by Oxman's Material-Based Design Computation [1], which was developed as a doctoral thesis at MIT and later developed in a body of research and built work. Its impact on the Rhinoceros™-Grasshopper™ ecosystem's subsequent development is traceable throughout the platform's research applications. Rutten's Grasshopper™, currently the most diffused design and research platform for the integration of AI in architectural and product design practice that, thanks to its software—the dataflow visual programming model and the component-based plugin system—has influenced the integration of AI technologies especially at the design scales described in previous chapters [2]. Kyratsis et al.'s CAD-based programming research at the University of Western Macedonia [3, 4] aims to demonstrate the possible integration of AI-generative tools within the most used design platforms for precision manufacturing applications, and provides a complementary perspective for the chapter's analysis between design and production.

Indeed, the chapter is organized as follows. Section 5.2 defines the conceptual framework to describe the technology-to-fabrication technology at disposal. Section 5.3 is a review concerning parametric design environments and their capacity to integrate AI applications. Section 5.4 focuses on AI-native creative platforms and their use and implementation in professional design practice. Section 5.5 delves into fabrication-integrated AI pipelines focusing on addictive and subtractive practices. Section 5.6 investigates multi-model AI workflow orchestration and API ecosystems. Section 5.7 focuses on the issues of design agency, platform dependence, and the political economy of AI design tools. Section 5.8 synthesizes benefits, limitations, and open research directions. Finally, Sect. 5.9 provides a conclusion bridging to Chap. 6.

5.2 The Design-To-Fabrication Technology Stack: A Conceptual Framework

In software engineering, a technology stack is the layered combination of programming languages, frameworks, databases, and infrastructure that all together define a software application and, in the context of this volume, it provides an analytical framework for understanding the platform infrastructure of AI-assisted design. In a software stack, each layer is interdependent of another and provides service to the one above while depending on the one below; therefore, a single layer is not self-sufficient and only from the interaction of each components the capabilities of the full stack emerge. This is very similar to a design-to-fabrication technology stack where geometric modelling capabilities, performance simulation engines, AI instruments, and manufacturing control system operate progressively to design and produce artefacts.

5.2.1 Layers of the Design-To-Fabrication Stack

The design-to-fabrication stack can be represented as composed of five layers, spanning from the more abstract to the more physical ones. First is the data layer that provides the representations characteristics—geometric data formats, building information models, geographic information systems, material property databases—where every specific design data is organized, structured, and stored. Second is the computational layer dedicated to the processing capabilities such as parametric geometry engines, finite element solvers, fluid dynamics simulators, and AI inference runtimes that are used to generate, transform, and evaluate the design solutions. Third is the interaction layer that serves to create the interfaces used by the human design to engage with the computation capabilities (graphical design environments, visual programming tools, natural language input systems, immersive visualization platforms, etc.). Fourth is the communication layer that is in charge of all the translation services—format converters, data exchange protocols, API interfaces—used to convey information between the different layers of the stack. Fifth, and last, is the fabrication layer dedicated to manufacturing capabilities—CNC machine controllers, additive manufacturing firmware, robotic motion planners—useful to translate all the different design data into physical form.

AI models function concurrently across several levels of this stack. For example, a diffusion model dedicated to architectural visualize can work at the computational layer, employing text input to generate image data; if integrated in a design environment, it can also operate at the interaction layer, when its outputs are displayed together with conventional CAD geometry; at the communication layer it can provide the translation of its outputs into three-dimensional geometry for BIM environments; finally, the possible fabrication of the elements extrapolated from its outputs involves the fabrication layer. To understand the impact in design workflows of a specific AI model, it is then fundamental to understand the precise stack level where it can be deployed and operate [5].

5.2.2 Bottlenecks and the Representation Gap

The representation gap—which is the mismatch between the outputs generated from AI models and the data formats required by conventional tolls in a design-to-fabrication stack—is one of the most reported processual bottlenecks in AI design literature. At various layer transitions, this gap appears in different ways. AI-generated point clouds or mesh geometries must be shown and assessed in design environments optimized for parametric solid geometry at the computation-to-interaction transition. At the interaction-to-communication transition, AI design concepts have the need to be converted from visualization-quality representations to quality models that can undergo performance analysis from an engineering perspective. Furthermore, engineering models must be converted into manufacturing instructions that the fabrication layer can follow throughout the communication-to-fabrication transition. These instructions include toolpath files, build orientation parameters, and support structure geometry.

To successfully conduct all these transitions, a certain amount of manual design intelligence (a practitioner that can understand and mediate among all the needed layer transition alongside the stack) or automated conversion software of sufficient sophistication to perform the translation reliably without human intervention is needed. Many of the current existing practices are based on the first option while the second one represents still one of the main technical challenges for a proficient implementation of AI in professional design contexts and workflows. The research areas mentioned in the previous chapters—fabrication-aware generative design systems, AI-to-BIM conversion tools, and CAD-native generative models—all tackle the same basic problem of stack-layer translation from various perspectives.

5.3 Parametric Design Environments and AI Integration

Parametric design software are environments where geometric models are synthetized as a function of adjustable parameters (often represented as nodes) where computational processes can be applied to these models using scripting or visual programming. This family of software, as already mentioned, are presently leading the research towards AI integration in the design field. Their role as integrators—platforms that can integrate AI inference capabilities with parametric geometry generation, performance simulation, fabrication preparation, and interactive design exploration within a single, cohesive workflow environment—is what makes them significant, not their AI capabilities per se, which are usually obtained through third-party plugins or API calls.

5.3.1 Rhinoceros and Grasshopper: The Research Platform of Record

McNeel's Rhinoceros 3D™, also thanks to its visual programming extension Grasshopper™ developed by David Rutten [2], is one of the leading research platforms for AI integration during the last decade. The dataflow visual programming model of Grasshopper™—where computational processes and generative parameters are expressed as nodes connected by wires, and data flows through graphs—offers a flexible framework to combine diverse computational capabilities spanning from the generation of parametric geometries, physical simulation, performance analysis, and AI inferences. All these operations can be reduced to Grasshopper components and connected through a single computational graph.

The dataflow visual programming model of Grasshopper™—in which computational processes are represented as nodes connected by wires, and data flows through the graph from inputs to outputs—provides a highly accessible framework for combining heterogeneous computational capabilities: parametric geometry generation, physical simulation, performance analysis, and increasingly AI inference, can all be represented as Grasshopper™ components and connected within a single computational graph.

Topology optimization (Millipede™, Karamba3D™), environmental performance simulation (Ladybug™ Tools, Honeybee™), machine learning inference (Lunchbox™ ML, various Python™-scripted components communicating with cloud AI APIs), and generative design (Octopus for multi-objective optimization, various GANs and diffusion models accessed through HTTP API calls) are all part of the large Grasshopper plugin ecosystem. This diversity implies that a software like Grasshopper™ can cover the design-to-fabrication stack: it generates parametric geometry and evaluates its environmental and structural performance through AI-driven form generation or optimization, and then create fabrication-ready output data for addictive manufacturing or CNC milling.

The main limitation behind the Grasshopper™ environment in terms of AI implementation is the same that, more generally, characterizes visual programming environments: the growing difficulty in managing computational complexity as the script and workflow get more complicated. For example, a Grasshopper™ definition that includes parametric geometry, finite element analysis, AI inference, and fabrication preparation can involve hundreds of different components with thousands of related connections/wires; such definition is not only difficult to read but also to modify and debug. This results in a tendency in accumulating a so-called technical debt—informal workarounds and undocumented dependencies—that makes them unstable if modification occurs. Therefore, the practical scalability of Grasshopper™-based AI workflows is directly impacted by the platform research challenge of developing higher level abstractions for AI-augmented parametric design, which encapsulate intricate AI-integration patterns as reusable components with clean interfaces [6].

In the precedent chapters, the importance of Grasshopper™ in AI design research is already documented. In Chap. 3, Chaillou's ArchiGAN [7] was developed and operationalized using a Python-Grasshopper integration while the topology optimization workflow discussed in Chap. 2 has already been proficiently deployed in Grasshopper™ through nTopology™ or Millipede™ plugins. Furthermore, Ladybug™ and Honeybee™ of Grasshopper™ have been crucial to building performance optimization workflows examined in Chap. 3, and Grasshopper™ has been the main parametric environment used in the urban layout optimization study covered in Chap. 4. One important aspect of the Grasshopper™ ecosystem's current importance to design AI research practice is the cross-scale, cross-domain uniformity of platform selection.

5.3.2 *Industrial CAD Platforms and Generative Design*

Some of the major industrial CAD platform—such as Autodesk™ Fusion™ 360, Siemens™ NX™, PTC™ Creo™, Dassault Systèmes™ CATIA™, and SolidWorks™—worked on the integration of generative design features inside their core interface, with the aim of making AI-based structural optimization accessible to engineers and manufacturing industries without the need for computational experts. These platform-integrated generative design features mainly function at the optimization modality mentioned in Chap. 1: the system creates structurally optimized geometry that satisfies the loads, boundary conditions, material, and manufacturing process constraints specified by the designer; the designer then chooses from the generated alternatives and develops the chosen configuration within the familiar CAD environment [8].

The research from Kyratsis et al. on CAD-based programming research [3, 4] offers an interesting perspective on the possible integration of AI technologies inside industrial CAD platforms that goes beyond the generative features of commercial platforms and focuses on the automation of CAD-based design and manufacturing processes through specific APIs. SolidWorks™ and Autodesk™ Platform Services API serve to showcase not only the full geometry generation and modification features, but also the possibility to export to external programming environments and allowing the development of AI-driven design automation systems that adhere to the parametric associativity standards of the CAD environment and function within its data model. For instance, Tzotzis et al. [9] used this approach to generate a G-code for CNC drilling operation starting from CAD model feature recognition, a procedure that generates manufacturing instructions without the need for human process planning interaction by using AI-based feature detection to power a rule-based process planning system via the SolidWorks API. This integration pattern—AI inference powering API-mediated CAD operations—represents a workable and instantly implementable method of design automation that doesn't call for the creation of new generative modelling tools. Its expansion to more complicated design tasks is a fruitful short-term research avenue.

5.3.3 *Oxman's Material-Based Design Computation: A Foundational Framework*

In her Material-Based Design Computation framework [1], firstly conceived during her professional studies at the Architectural Association in London and later developed as a doctoral thesis at the MIT's Media Lab in the Mediated Matter research group, Neri Oxman proposed a technical and conceptual network for computational design that, undoubtedly, had an impact on parametric design platform and its AI integration. In Oxman's theorization, material properties—intended as the mechanical, thermal, optical, and biological behaviours of physical matter—must be considered as primary in a computational design process, influencing geometry generation from the beginning and not just incorporated a posteriori.

To achieve that, there is the need to develop computational design frameworks that could represent these material properties in the form of continuous spatial field that can influence geometry generation, and not as simple property assignments to already defined geometric elements. The multimaterial additive manufacturing capabilities that are now commercially available in platforms like Stratasys™ PolyJet technology and Carbon's CLIP process were foreseen by Oxman's voxel-based material modelling approaches, which represent three-dimensional material property distributions as volumetric data and are used to drive additive fabrication processes that deposit different materials at different spatial locations.

The above-mentioned research at Mediated matter, that includes project such as Wanderers, Mushtari, and the Silk Pavilion, showed an integration of biological growth algorithms, finite element structural optimization, and additive fabrication through a material-driven computational design environments that still no commercial platform can allow [10].

Oxman's framework is important for this chapter not only as a technical contribution but also as a design philosophy. It shows that the most effective way to incorporate AI into design platforms is to rethink the platform itself around the capabilities of the AI, rather than adding it as a feature to an already-existing workflow. In this perspective, the Mediated Matter research platform was design from scratch with the aim to support material-intelligent generative design as a foundational idea. One of the most significant implications of the platform literature for the design of future AI design tools is this lesson: platform architecture shapes the character of the AI design practice it enables, and the most productive AI design platforms may require architectural rethinking rather than feature addition.

5.4 AI-Native Creative Platforms and Their Relationship to Design Practice

A new class of design tools has emerged that sits alongside, rather than inside, the parametric design environments and industrial CAD platforms discussed in the previous section. This is due to the quick commercial development of AI-native creative platforms, which are tools created from the ground up to offer direct access to generative AI capabilities through interfaces targeted towards creative rather than engineering users.

Such platforms, basically image generation services, three-dimensional forms generation APIs, and large language model design assistants, are known for a high level of accessibility and creative speed but not for the parametric associativity and technical precision that professional workflows required, therefore their implementation in the professional field is to be considered productive but, at the same, limited.

5.4.1 Image Diffusion Platforms in Design Practice

Since their appearance in the market, text-to-image diffusion platforms—among the most the famous Midjourney™, Stable Diffusion™ [11], Adobe Firefly™, DALL·E 2™—have been widely used as early-stage ideation tools, for the peculiarity to quickly generate photorealistic outcome images starting from natural language descriptions (prompts). Concerning their practical integration into professional workflows, the situation it is variable and depends on the discipline under consideration. For example, in architectural design, diffusion models-generated images are used mainly for competition visualization and as a client communication tool at the concept/idea development stage, considering their ability to rapidly generate renderings of design directions before consuming time in the creation of articulated three-dimensional models that could guarantee the same level of rendering quality [12]. In the field of product design, without the need for physical prototype, diffusion tools are utilized for material and finish exploration, producing visualizations of the same shape in several material treatments for customer evaluation. Diffusion-generated aerial perspective imagery and street-level visualizations are utilized in urban design to convey recommendations for urban transformation in ways that non-specialist community stakeholders can understand.

Nevertheless, as discussed in Chaps. 3 and 4, the current limitation in the implementation of diffusion platforms in professional design/architecture must be clearly understood. Undoubtedly, the main limitation stands on the absence of the geometric coherence since most of the diffusion-generated images, even though showcasing visual appealing spatial configurations of buildings, objects, urban spaces, etc., present a considerable level of spatial incoherence that becomes apparent under professional scrutiny. Bommasani et al. [13] argue that the main reason is a consequence of the two-dimensional nature of the training datasets for large foundation modes: family of models that can learn the statistical regularities and repetition in form appearance but they miss the development of a concrete three-dimensional understanding that it is one of the essential conditions behind design judgement. Furthermore, for the AI-native platform ecosystem, the creation of three-dimensional coherent design AI—models that produce spatially consistent representations rather than visually plausible images—remains a major unresolved problem.

Another limitation that requires proper consideration in image diffusion platforms is the intellectual property issue. Diffusion models are trained on a large corpus of images acquired from the Internet—that often includes also copyrighted images, drawings, project documentation, etc.—and in many cases they could implement in their outcomes some stylistic elements from a specific practice without acknowledging it. Indeed, the legal status of AI-generated images in professional context is still under discussion, and the creation of specific legal and professional frameworks for AI images is an addressed challenge that impact further professional development of these tools. Both of these limitations—the absence of three-dimensional spatial coherence and the inconsistency of stylistic output across views—have motivated the development of architecture-specific AI-native platforms designed to address the professional discipline's requirements more directly than general-purpose image diffusion instruments. FORMAS.AI [28], developed by architect and professor Carlos Bañón and entrepreneur Yiping Goh, is an interesting example in this category.

The platform advertises itself as an AI orchestrator design studio where *form follows you*, and integrates a Freeform canvas for sketch-based ideation, an Agent 2.0 generation engine capable of interpreting gestural and textual inputs simultaneously, and a Style Memory system that maintains formal and material consistency across multiple views and iterations of the same project, addressing one of the most frequently cited limitations of general-purpose diffusion tools in architectural practice. Furthermore, the Viewpoint studio module allows the generation of coordinated perspective images from a single design concept—elevations, sections, aerial views, street-level perspectives—while maintaining a spatial coherence without the need for specific prompt complexities or conditioning procedures that other platforms would require. Furthermore, other features include the ability to create video walkthroughs from still design images, export point clouds for further geometric processing, and produce output in 2K or 4K quality for professional presentations and competition submissions.

Currently, the platform is under adoption by the students of many worldwide-known institutions such as Columbia University, Cornell AAP, the Architectural Association, IAAC, the University of Pennsylvania, and many professional studios, confirming the market needs form architecture-specific AI-native platforms that can be used in high-skilled professional contexts.

5.4.2 *Large Language Models as Design Assistants*

The integration of large language models (LLMs)—including GPT-4™ [15], Claude™, and their successors—within design workflows represents a qualitatively different kind of AI-native platform from image generation tools. LLMs operate primarily at the text and code levels rather than the image level, and their most practically significant contributions to design are in code generation (the automated writing of parametric design scripts, Grasshopper™ component definitions, and CAD API calls from natural language specifications), design brief interpretation (the translation of client requirements and regulatory specifications into structured design program information), and documentation generation (the automated production of design reports, specification documents, and regulatory submissions from design model data).

The capacity of LLMs to generate functional parametric design code from natural language specifications is particularly significant for the integration of AI within parametric design environments. A designer who can specify a desired geometric operation in natural language—'generate a grid of columns with variable heights responding to a user-defined surface'—and receive a functional Grasshopper™ component definition or Python™ script implementing that operation has access to parametric design automation without requiring the programming expertise that conventional scripting demands. Achiam et al. [15] support the evidence that GPT-4™ can generate functional Python™ code for complex computational tasks, including geometry generation and data processing operations, from natural language specifications with a sufficient reliability for practical use in workflows where generated code is reviewed and tested before deployment. The extension of this capability to domain-specific design scripting environments—Grasshopper™, Dynamo™, the Revit™ API—is an active area of commercial development and academic research.

5.4.3 Three-Dimensional Generation Platforms

The most direct AI-native platform contribution to the gap between concept generation and geometric modelling, which has been identified throughout the previous chapters as a primary workflow bottleneck, is the creation of commercially available three-dimensional shape generation platforms—services that generate three-dimensional models from text prompts, image references, or sketch inputs.

Shap-E™ [16], Point-E™, Meshy™ AI, and other three-dimensional diffusion services have the ability to produce three-dimensional geometries directly from text or image conditions and can produce mesh outputs that theoretically can be directly sent to other design tools for further refinement.

Nevertheless, the practical use of such outcomes in professional workflows is still limited by the quality gap analysed in the precedent chapter of this book: usually, the generated geometries suffer from surface continuity, topological regularity, and dimensional precision that are needed features for engineering analysis and manufacturing processes. Zeng et al.'s LION [17], and other latent point diffusion approaches have shown sensible improvements in output quality, even though the gap between AI-generated geometry and the manufacturing-ready models required by professional practice still remain problematic. The timeline for practically deployable three-dimensional generation in professional design contexts is getting shorter, but the trajectory of improvement is quick—three-dimensional generation platform quality has improved more in the last three years than image generation quality did in the five years after the introduction of the GAN.

5.5 Fabrication-Integrated AI Pipelines

In the design-to-fabrication stack, the fabrication layer—namely all the systems that can translate digital design geometry into ready-to-manufacture instruction that can be interpreted through CNC machines, additive manufacturing equipment, or robotic fabrication systems—is definitely the domain where AI integration has the most direct impact on product quality and manufacturing productivity. More specifically, at the fabrication layer stage, AI models do not generate forms but rather optimize the process of translation from form to physical material, while compressing the cost and time behind manufacturing preparation and improving the dimensional consistency of the fabricated outcomes.

5.5.1 AI-Assisted Process Planning and Toolpath Generation

Translating three-dimensional CAD geometry into CNC manufacturing instruction always required professional engineering know-how addressing the identification of the right machine features from the original CAD models; the sequencing and selection of the needed machine operations; the specification of cutting parameters and tools; and the final generation of toolpaths. Some AI approaches that showed practical significance in this field emerged during the last years.

Both rule-based systems and machine learning classifiers trained on labelled CAD model datasets have been used to demonstrate feature recognition from CAD models, which is the automatic identification of geometric features (holes, pockets, fillets, bosses) and their association with suitable machining strategies [18]. The integration of AI feature recognition in CAD platforms APIs allows the establishing of CAPP systems (Computer-Aided Process Planning) able to generate machining strategies directly from CAD model inputs without manual intervention as demonstrated for example by Fountas et al. [19] concerning CNC sculptured surface machining optimization. Particularly for designs with intricate geometric elements that would take hours for skilled process planners to manually plan, these technologies shorten process planning times and increase the uniformity of machining parameter choices.

Regarding cutting parameter optimization—predicting optimal spindle speed, feed rate, and depth of cut for a given material, tool, and surface geometry combination—neural network-based models are usually trained on experimental machining data and implemented in CAM software environments with the aim to provide automated parameter recommendations. The practical productivity implication of these optimization systems—reduced setup time, improved surface quality, and extended tool life relative to empirically selected parameters—provides a clear commercial justification for their deployment in production manufacturing environments.

5.5.2 AI for Additive Manufacturing Process Control

A specific class of AI application regarding the fabrication layer is introduced by additive manufacturing process, respecting their characteristic of layer-by-layer material deposition: the potential for real-time process monitoring and adaptive control, the sensitivity of component quality to process parameters that change throughout the build volume, and the creation of intricate internal geometries that require destructive testing for post-build inspection. All three of these traits have benefitted greatly from the application of AI.

Current neural network models for the prediction of additive manufacturing process outcomes—e.g., roughness, dimensional accuracy, porosity, and mechanical properties—are trained on experimental datasets and showed a considerable precision accuracy that can be used in parameter optimization through pre-built simulation environments [20].

By enabling the quick assessment of different parameter sets for a specific shape prior to committing to a physical build, these surrogate models lessen the costly empirical parameter optimization that is currently necessary for the development of additive manufacturing processes. For FDM, SLS, and metal powder bed fusion processes, computer vision systems for real-time in-process monitoring of additive manufacturing builds have been demonstrated. These systems are trained to detect layer deposition anomalies, warping, cracking, and other defects from camera imagery of the build surface. This enables adaptive process control that modifies deposition parameters in response to detected anomalies rather than discovering build failures after completion [21].

A proficient integration of AI design generation with AI-controlled additive manufacturing is the closest thing to the fully design-to-fabrication pipeline discussed in Chap. 1 and one of the main aspirations for AI-assisted design and prototyping. Ren et al. [20], propose a CAD-based framework that reconstructs editable, CAD compatible solid models from the mesh geometry typically output by topology optimization and generative design processes—closing a longstanding representation gap between algorithmically generated design output and downstream manufacturing preparation. Theoretically, a closed-loop design-to-fabrication pipeline in which AI models engage at every stage from concept development to actual production is finished by combining this framework with generative design tools and real-time process monitoring.

5.5.3 Robotic Fabrication and AI-Controlled Assembly

The field of robotic fabrication—concerning the use of industrial robots activated through CAD/CAM interfaces to perform complex operations such as wire arc additive manufacturing, concrete three-dimensional printing, timber assembly, and composite layup—is definitely one of the sub-fields in the fabrication layer where AI deployment is running at a fast pace. If before, specialist robotic programming expertise was required for robotic fabrication systems related to complex geometries; now, AI-assisted toolpath generation in robotic platforms—namely, automate systems that can convert any design geometry into robotic motion without the need for programming skills—represents an area where both academic and professional research is focusing [22].

In robotic assembly planning, the use of reinforcement learning—using simulated interaction with a physics model of the assembly process, agents are trained to find effective assembly sequences for intricate structural components—showed interesting capacity to find the best assembly strategy that, in most cases, the human counterpart would not easily discover, especially for structures with constrained access geometries or multi-robot coordination requirements [23].

Furthermore, the integration of these RL-based assembly planners in parametric design environment—giving designers the possibility to specify all the needed assembly requirements starting from the design brief phase, and to receive robot-feasible assembly procedures as an output—is considered today an unprecedented possibility to close the design-to-fabrication loop in architectural and structural design.

5.6 API Ecosystems and the Orchestration of Multi-Model AI Workflows

As already said, all the platforms analysed and presenting in the preceding sections of this volume do not operate as isolated systems. For example, in relation to professional design workflows, they are usually merged and orchestrated through application programming interfaces (APIs) that allow data and computation to pass through components standing at the different layers of the design-to-fabrication stack. Nowadays, the possibility to design effective API ecosystems—addressing among the others the specification of data formats, interaction protocols, and service interfaces that enable productive interoperability among parametric design tools, AI inference services, performance simulation engines, and fabrication control systems—is considered a critical point on the utility of AI application in professional design contexts.

5.6.1 REST APIs and Cloud AI Services

The wide adoption of REST API architectures in relation to cloud computing services has created a standard interpolation pattern to integrate cloud-hosted AI inference services within design platform workflows: practically the process can be synthetized as follows. A design tool sends an HTTP request to an AI inference API, where the design data pass as a JSON payload; after receiving the request, the API activates inference on a cloud-hosted hardware and then sends back the results in a specific format; then, the design tools acquire and process the returned results to integrate them within their design environment. This pattern is widespread because it is simple, platform-independent, and does not require a local installation of the AI model in order to make it more accessible to practitioners that do not possess a dedicated AI infrastructure.

The limitation of REST API integration for real-time design feedback—the latency introduced by network round-trips and cloud inference queuing—is significant for applications where continuous, interactive AI feedback is required, such as real-time structural performance assessment during parametric form exploration.

To address this issue, local inference deployment of smaller and purposed-trained AI models is used even though reducing the model capability and demanding considerable computational resources. In this perspective, an important research direction is the one focused on the development of efficient and locally deployable AI models for design purposes and exploiting the maximum capacity of these tools [24].

5.6.2 Multi-Model Pipeline Orchestration

The sequential multi-modality workflows analysed during the previous chapters—e.g., diffusion model concept generation feeding VAE latent space navigation feeding topology optimization—and explained in Chap. 2 through the footwear design case, and in Chap. 3, at the architectural scale, through the ArchiGAN floor plan generation—rely on the coordination and orchestration of multiple AI models that operate through different data representation within a coherent workflow sequence.

To construct and make such a complex pipeline efficient, it is not enough to have dedicated API integration at each model boundary; it is also necessary to design a data transformation logic that facilitates the transition of the output representation from a model to another.

Multi-modal representation learning—the ability of a single model to function across text, image, and three-dimensional geometry modalities simultaneously—was identified by Bommasani et al. [13] as a crucial path for lowering the representation conversion burden in multi-model design pipelines, in other words, the creation of a multi-modal foundation model that could span among text descriptions, architectural imagery, and three-dimensional geometry generation through a unified representation framework to reduce the friction that actually characterizes multi-modal design workflows. The trajectory of multi-modal AI development, as demonstrated by the text and image capabilities of GPT-4™ [15] and by emerging unified vision-language-geometry models in the research literature, indicates that this limitation is likely to be gradually addressed within the medium-term future, even though such models do not currently exist at the capability level required for professional design use.

5.6.3 Open-Source Ecosystems and the Democratization of AI Design

A critical role in democratizing the access to AI technologies has been played by the existing AI ecosystem, including frameworks like PyTorch™, TensorFlow™, and JAX™; model repositories such as Hugging Face™ and GitHub™; and the growing library of open-source implementations of AI design research. The possibility to use pre-trained model for architectural design and floor plan generation, prediction of building performances, urban morphology analysis, and three-dimensional shape generation, hosted in open-source repositories empowered the global community of researchers and practitioners to benefit from AI design capabilities without the need for expensive commercial licensing.

In 2022, Stability AI, through the Stable Diffusion model release, created a milestone in the democratization of open-source AI in design practice: through the release of both model weights and inference code under a shared license, the company allowed a community of developers to use it to create design-oriented applications, architectural fine-tuned variants, and integrated design workflow tools at a pace that no private company sustains. Open-source releases of models and datasets, such as ShapeNet™ [25], OSMnx™ [26], Urbanity™ [27], and various floor plan generation model implementations, have also benefitted the design AI research literature. These releases have accelerated research progress and allowed cross-institutional collaboration of a scope not previously possible in the field.

5.7 Platform Dependency, Design Agency, and the Political Economy of AI Design Tools

Undeniably, the platform landscape presented in the current chapter is not a mere technical infrastructure, but rather a powerful commercial ecosystem where software houses and companies play a crucial role in defining the design capabilities available to practitioners, the terms under which this access is granted, and the collection of data generated through their use. The division of power, profit, and risk among the platform providers, the design firms using their tools, and the communities whose physical environments are affected by the designs those tools assist produce is known as the political economy of AI design platforms, and it merits specific critical attention.

5.7.1 Vendor Lock-In and the Autonomy of Design Practice

The autonomy of design practices and educational institutions is limited by the concentration of AI design capabilities in the hands of a small number of commercial platforms (Autodesk™, Siemens™, Dassault Systèmes™ in industrial CAD;

Adobe™, Stability™ AI, Midjourney™ in creative AI; Nvidia™ in AI hardware) with the risk of generating a potential vendor lock-in.

In more professional terms, a design firm that built all its AI-based workflow on a proprietary platform is dependent in relation to continued availability, pricing, and capability development, while a university that teaches AI design only through a commercial platform—considering that the license could be expensive or limit the students' access—is extremely vulnerable in terms of quality and continuity of education. For this reason, the reviewed (Sect. 6.3) open-source AI ecosystem is an essential counterpart to limit vendor lock-in and enhance the construction of more inclusive and democratic AI workflows based on open-source solutions that bring with them no license dependencies.

The practical limitation of open-source AI tools in professional design contexts—the greater effort required to deploy, maintain, and support open-source tools relative to commercial products—means that the choice between open-source and commercial AI platforms involves genuine trade-offs between autonomy and convenience that design organizations must navigate according to their specific strategic circumstances.

5.7.2 *Data, Training, and the Ownership of Design Intelligence*

The quality and relevance of data used to create training datasets is a fundamental point to properly evaluate the capacity of AI design platforms. Regarding professional design contexts, one of the most relevant trainings is through the professional own design work—series of built projects, design drawings, fabrication records, and performance data—that are already part of the studio archive and cover a broad dimension of typologies, programmatic functions, and aesthetic features. Nevertheless, the way in which commercial platforms benefit from this richness of data is not always transparent, and the question concerning who really owns the design intelligence incorporated in AI platforms—trained on high-quality professional materials—is currently an unresolved issue under study from different research perspectives. An institutional and legal challenge of great practical significance is the creation of data governance frameworks for the professional design disciplines. These frameworks should outline the circumstances under which data from design firms can be used for AI training, guarantee sufficient compensation for data contributors, and safeguard the intellectual property of design practitioners. Its resolution will shape the terms on which the AI design platform ecosystem develops and the distribution of benefit between platform providers and the design professions that generate the data on which AI design capabilities depend.

5.8 Benefits, Limitations, and Open Research Directions

5.8.1 Demonstrated Benefits

Several categories of benefit deriving from the implementation of AI techniques in design practice and research have been delivered through the platform landscape analysed in this chapter. In regard to parametric design environments, Grasshopper—and its related ecosystem—empowered a global community of designers and architects to get familiar with AI design capabilities with the aim of deploying them through a full range of spatial scales and design disciplines, with a level of freedom and workflows diversity that commercial platform cannot guarantee. The research from Kyratsis and his group [3, 4, 9] addressing CAD-based programming research showed how systematic API-level integration of AI within industrial CAD platforms can enhance quantifiable benefits—such as reduced process planning time, improved parameter optimization, and consistent G-code generation—in manufacturing productivity and related contexts.

Regarding fabrication integrated pipelines, AI-assisted addictive manufacturing process control proved considerable quality improvements—reduced defect rates, improved dimensional accuracy, and extended equipment utilization—that validate commercial deployment in production AM environments [14, 20]. Finally, in AI-native creative platforms, diffusion models have accelerated the early concept development phase in design and architectural practice, allowing the investigation of a broad range of design option in shorter timeframes while improving the standard of client communication during the concept development phase [12].

5.8.2 Principal Limitations

Currently, the most consistent limitation that can be found in design literature is the already mentioned representation gap—the difference between AI outputs and the data formats required by the tools involved in a design-to-fabrication stack—and that still represents the primary source of friction that restricts AI design integration in professional contexts. This gap is not entirely filled by any platform tier now in use; each one necessitates human-mediated translation at crucial stack-layer transitions, which limits the productivity boost that AI automation can achieve.

Another issue is platform fragmentation—the spread of inconsistent workflow patterns, API specifications, and data formats throughout the ecosystem of AI design platforms—that creates the need for surplus costs in design practices to combine the outputs produced by multiple platforms. Furthermore, on this topic, the absence of precise interoperability standards for AI data exchange—a design-domain equivalent of the open standards that enable interoperability in adjacent digital infrastructure domains—is a major structural constraint of the existing platform environment that has not yet been sufficiently addressed by the design technology community [5].

Finally, the political economy limitations—vendor lock-in, data ownership ambiguity, and the concentration of platform power—limit the autonomy of design practices and educational institutions generating some long-term implications for further developments of AI-assisted design practices.

5.8.3 Open Research Directions

For the platform and ecosystem aspects of AI-assisted design, three research directions are particularly significant. First, the creation of standards for AI design data exchange to guarantee format interoperability—e.g., specifications for the representation and transfer of AI-generated geometric data, performance predictions, and design rationale between platform components—as a precondition to address platform fragmentation and the development of productive multi-model workflows without the current need for expensive integration costs. Such challenge happens both at the professional and academic levels, and requires a collective engagement between academic researchers, commercial platform developers, and professional design entities.

Second, the creation of platform architectures that incorporate AI capabilities into design tools in ways that enhance rather than replace designer intelligence—interfaces that preserve the designer's comprehension of the relationship between design choices and their effects—make AI's generative logic readable, and encourage designers to exercise judgement at crucial decision points. In this perspective, Oxman's Material-Based Design Computation framework [1] suggests a model that can be taken into account to understand how platforms can amplify rather than automate design intelligence.

Third is the development of specific frameworks for data governance in professional disciplines. Such frameworks should address the ownership of training data, the attribution of AI-generated design outputs, and the equitable distribution of benefit between platform providers and data-contributing design practitioners, and improve the existing institutional, legal, and ethical dimensions that go beyond the mere technical domain of AI research. This holistic perspective is mandatory for a healthy and sustainable development of AI-based design platform.

5.9 Conclusions

The platform and ecosystem infrastructure that facilitate the development, distribution, integration, and application of AI design skills across the three geographic scales covered in this volume's earlier chapters have been investigated in this chapter.

Three main categories of design-to-fabrication technology have been described: visual programming platforms and parametric environments, embedding AI potentialities into design familiar workflows; AI-native creative platforms that give the user direct access to all the model generative capabilities; and AI pipelines that focus on the translation of generative outputs into manufacturable objects.

Examined as aspects of the platform question that go beyond its purely technical aspects are the API ecosystems that link these tiers, the educational implications of the platform landscape for design curricula, and the political economy of AI design tool development and deployment.

The discussion was structured around three case studies drawn from existing literature. First is Neri Oxman's Material-Based Design Computation [1] that serves as a foundational model for platform that has the aim to amplify intelligence rather than displace it, and since its publication had a strong influence on AI in design research. The Rhinoceros™-Grasshopper™ ecosystem [2], considered as the main planform for a research-level integration of AI in architecture and product design. Finally, Kyratsis et al.'s work on CAD-based programming research [3, 4] testifies the productivity boost deriving from a systematic AI integration in industrial CAD/CAM workflows, and represents a design domain that complements and enhances both architectural and product design horizons.

Chapter 6 aims to take the theoretical reflections in this volume a step further, and to propose a way to overcome 'AI-to-manufacture gap'—that is the distance between the outcomes produced by AI generative models and the level of readiness in geometries required by manufacturing processes—using an original case study from the authors that implement different family models towards digital fabrication.

References

1. Oxman N (2010) Material-based design computation (Doctoral dissertation, Massachusetts Institute of Technology)
2. Rutten D (2013) Grasshopper: algorithmic modeling for Rhino. McNeel Europe, Barcelona. https://www.grasshopper3d.com
3. Kyratsis P, Tzotzis A, Davim JP (2025) Automated assembly of mechanical systems: a case study. In: CAD-based programming for design and manufacturing, pp 53–68. Cham: Springer Nature Switzerland
4. Kyratsis P, Tzotzis A, Davim JP (2025) A Review on CAD-based programming for design and manufacturing. CAD-based programming for design and manufacturing, 1–26
5. Eastman CM (2011) BIM handbook: a guide to building information modeling for owners, managers, designers, engineers and contractors. John Wiley & Sons
6. Menges A, Ahlquist S (2011) Computational design thinking: computation design thinking. John Wiley & Sons
7. Chaillou S (2019) Archigan: a generative stack for apartment building design. NVIDIA Corporation
8. Buonamici F, Carfagni M, Furferi R, Volpe Y, Governi L (2021) Generative design: an explorative study. Comput Aided Design Appl 18(1)
9. Tzotzis A, Manavis A, Efkolidis N, Kyratsis P (2021) CAD-based automated g-code generation for drilling operations. Int J Modern Manuf Technol (IJMMT), 13
10. Oxman N (2016) Age of entanglement. J Design Sci
11. Rombach R, Blattmann A, Lorenz D, Esser P, Ommer B (2022) High-resolution image synthesis with latent diffusion models. In: Proceedings of the IEEE/CVF conference on computer vision

and pattern recognition (CVPR), New Orleans, pp 10684–10695. https://doi.org/10.1109/CVPR52688.2022.01042
12. del Campo M (2022) Deep house—datasets, estrangement, and the problem of the new. Archit Intell 1:12. https://doi.org/10.1007/s44223-022-00013-w
13. Bommasani R, Hudson DA, Aditi E, et al (2021) On the opportunities and risks of foundation models. arXiv:2108.07258
14. Markopoulos AP, Georgiopoulos S, Manolakos DE (2016) On the use of back propagation and radial basis function neural networks in surface roughness prediction. J Ind Eng Int 12:389–400. https://doi.org/10.1007/s40092-016-0146-x
15. Achiam J et al (OpenAI) (2023) GPT-4 technical report. arXiv:2303.08774
16. Jun H, Nichol A (2023) Shap-E: generating conditional 3D implicit functions. arXiv:2305.02463
17. Vahdat A, Williams F, Gojcic Z, Litany O, Fidler S, Kreis K (2022) Lion: Latent point diffusion models for 3d shape generation. Adv Neural Inf Process Syst 35:10021–10039
18. Stavropoulos P, Tzimanis K, Souflas T, Bikas H (2022) Knowledge-based manufacturability assessment for optimization of additive manufacturing processes based on automated feature recognition from CAD models. Int J Adv Manuf Technol 122:993–1007. https://doi.org/10.1007/s00170-022-09948-w
19. Fountas NA, Vaxevanidis NM, Stergiou CI, Benhadj-Djilali R (2019) Development of a software-aided environment for optimizing CNC sculptured surface machining. Int J Adv Manuf Technol 104:1–20
20. Ren H, Xia B, Wang W, Chen X, Liu Y, Zhou P, Du J (2025) Automated CAD model reconstruction of topology optimization result. Comput Methods Appl Mech Eng 435, 117673. https://doi.org/10.1016/j.cma.2024.117673
21. Roy RB, Mishra D, Pal SK et al (2020) Digital twin: current scenario and a case study on a manufacturing process. Int J Adv Manuf Technol 107:3691–3714. https://doi.org/10.1007/s00170-020-05306-w
22. Rahman MA, Felix IA, Shahid U, Michaelis JE (2024) Pathwise: an ai-assisted teacher authoring tool for creating custom robot-assisted learning activities. In: Companion of the 2024 ACM/IEEE international conference on human-robot interaction, pp 88–90
23. Wang THJ, Zheng J, Ma P, Du Y, Kim B, Spielberg A, ... Rus D (2023) Diffusebot: Breeding soft robots with physics-augmented generative diffusion models. Adv Neural Inf Process Syst 36:44398–44423
24. Mishra A (2024) Scalable AI and design patterns: design, develop, and deploy scalable AI solutions. Springer Nature
25. Chang AX, Funkhouser T, Guibas L et al (2015) ShapeNet: an information-rich 3D model repository. arXiv:1512.03012
26. Boeing G (2017) OSMnx: new methods for acquiring, constructing, analyzing, and visualizing complex street networks. Comput Environ Urban Syst 65:126–139. https://doi.org/10.1016/j.compenvurbsys.2017.05.004
27. Yap W, Stouffs R, Biljecki F (2023) Urbanity: automated modelling and analysis of multidimensional networks in cities. npj Urban Sustain 3(1):45
28. FORMAS.AI (2025) FORMAS.AI: AI orchestrator design studio for architecture and interior design. Beta 3.0. https://www.formas.ai Accessed May 2025

Chapter 6
From AI Generation to Digital Prototyping: A Methodological Framework

Abstract The theoretical pillars, scale-specific analyses, and platform ecosystems of AI-assisted design and digital prototyping have been established in the previous chapters of this volume. The Five-Phase AI-to-Fabrication Design Pipeline, which is presented here as an original contribution for practitioners looking to incorporate AI generative tools into design-to-manufacture workflows in an organized, technically demanding, and educationally transferable manner, is a unified methodological framework that synthesizes these contributions. The chapter aims to addresses the AI-to-manufacture gap that has been identified as the main bottleneck throughout this volume, that is the existing distance between the outputs of AI generative models and their geometric readiness required to directly undergo manufacturing process. The Five-Phase framework, divided into five parts—Brief Definition, AI-Assisted Ideation, Geometric Reconstruction, Prototyping and Iteration, and Test and Evaluation—is explained covering its full methodological detail, and its operative logic is demonstrated through an original case study where an AI-generated backpack model is extracted from a diffusion-model generation and analysed through mesh validation and geometric correction to physical fabrication via Fused Filament Fabrication (FFF) using Creality™ K1 Max printer with NEEMA3D™ PLA:EVO filament. All of the fabrication process was documented—in terms of mesh topology, dimensional corrections, slicer configuration, print time, and post-processing effort—to identify where actually the AI-to-manufacture gap happens and how the Five-Phase framework can address it. The conclusion of the chapter offers a perspective on how AI capabilities—multi-agent design pipelines, geometry-native generative models, and fabrication-aware diffusion architectures—are likely to reshape the framework's operative requirements in the near-term future.

Keywords AI-to-manufacture gap · Five-phase framework · Digital prototyping · Diffusion models · Fused filament fabrication · Mesh validation · Geometric reconstruction · Design methodology · AI-assisted design · Backpack

P. Kyratsis et al., *AI Models for Design and Digital Prototyping*,
Manufacturing and Surface Engineering, https://doi.org/10.1007/978-3-032-35144-9_6

6.1 Introduction

The previous chapters of this volume have described the different spatial scales and platform environments which currently characterize AI-assigned design and prototyping. In Chap. 1, the theoretical and taxonomic foundations of the field were established, while three different operative modalities were described: optimization, augmentation, and generation, followed by a six-category taxonomy where AI models were divided according to different design tasks. Chapters 2, 3, and 4 were used to situate this framework to different growing-complexity scales (the object, the building, and the urban scale), to assess the different AI capabilities, workflow patterns, and specific human-AI cooperation model for each of those scales. Finally, in Chap. 5, the main platforms and ecosystems used to develop, spread, and integrate AI into professional practice were presented, spanning from visual programming pieces of software like Grasshopper™ to multi-agent design pipelines. One of the most recurrent challenges across all the discussed design scales was undoubtedly one: the AI-to-manufacture gap.

Introduced in Chap. 1—and encountered at all scales—the AI-to-manufacture gap is the existing distance between AI geometric representations from the geometry readiness that physical manufacturing processes require. In fact, if AI models can generate several types of output formats—such as point cloud, triangle mesh, neural implicit surface, and image representations—manufacturing processes work on watertight solid geometry grounded on well-defined wall thickness, coherent surface normal, accuracy in dimensions, and absence of topological errors like non-manifold edges and/or self-intersecting surfaces. As argued by Edwards et al. [1] for sketch-to-prototype pipelines, by Faruqi et al. [2] for AI-stylized 3D models, and by Li et al. [3] for domain-fine-tuned component generation, this translation between the two systems is not automatic and requires a decisive mediation from a human counterpart. Considering this, this chapter attempts to address this gap through the proposal, elaboration, and hands-on validation of a design framework that could help designers to orientate.

Therefore, the Five-Phase AI-to-Fabrication Design Pipeline represents the chapter's original methodological contribution, which aims not to focus on a single AI model or fabrication process but to be a methodology—a structured series of human and AI tasks with clear validation requirements and transition criteria at each phase boundary, intended to be adaptable across the variety of design scales, AI model families, and fabrication methods covered in the previous chapters. Indeed, the pipeline is divided in five phases: Phase 1, Brief Definition; Phase 2, AI-Assisted Ideation; Phase 3, Geometric Reconstruction; Phase 4, Prototyping and Iteration; and Phase 5, Test and Evaluation. The framework does not focus on a single scale or generative platform, meaning that Phase 2 can support the entire spectrum of text-to-image, image-to-image, and text-to-3D AI tools discussed in the earlier chapters without necessitating changes to the surrounding phases. This is applicable to product objects, architectural elements, and urban furniture.

To validate the operative logic of this framework, the original authors' case study is presented: the design and fabrication of a backpack using AI diffusion tools for its generative and Fused Filament Fabrication (FFF) on a Creality™ K1 Max printer for its production. The selection of a backpack as a design typology comes from several reasons: it blends together organic surface complexity—e.g., curved body panels, pocket intersections, strap attachment geometry—and specific functional and dimensional requirements coming from human body proportions. Therefore, it creates an interesting tension between AI generation—mostly based on aesthetic categories—and the constraints deriving from its functionality and essence of being an everyday consumer product design. The case study details every step of the process, from design brief to mesh validation, dimensional correction, slicer setup, and physical production. It also notes the particular geometric conditions that arise at each phase boundary and the human intervention needed to address them.

In software engineering, a technology stack is the layered combination of programming languages, frameworks, databases, and infrastructure that all together define a software application and, in the context of this volume, it provides an analytical framework for understanding the platform infrastructure of AI-assisted design. In a software stack, each layer is interdependent with another and provides service to the one above while depending from the one below; therefore, a single layer is not self-sufficient and only from the interaction of each components the capabilities of the full stack emerge. This is very similar to a design-to-fabrication technology stack where geometric modelling capabilities, performance simulation engines, AI instruments, and manufacturing control system operate progressively to design and produce artefacts.

6.2 The AI-To-Manufacture Gap: A Cross-Scale Synthesis

To determine precisely what the framework is intended to address and why it is necessary, it is helpful to organize the evidence for the AI-to-manufacture gap found across the previous chapters—and the existing literature—before presenting the methodological framework.

6.2.1 *The Gap's Structural Origins*

What is important to understand is that the AI-to-manufacture gap is not something related to the existing AI systems—and that will disappear as the tools evolve—but it is rather a deep discrepancy between the domains in which AI training and manufacturing requirements operate. Whether we refer to image diffusion models, GAN-based shape generators, VAEs, or text-to-3D systems, all the existing AI generative models undergo a training based on visual and geometric data that privilege appearance and aesthetics rather than manufacturability. For example, the training datasets of models like Stable Diffusion™, Midjourney™, or DALL·E™ 3 [4] do not include metadata concerning wall thickness, surface normal orientation, material behaviour, or manufacturing process constraints, while three-dimensional generative models like Shap-E™ [5] and LION™ [6] are trained through mesh and point cloud datasets that include geometry data but, again, no fabrication process physics. As a consequence—widely document in the literature [1–3]—all the AI-based outputs are plausible, under an visual perspective, but incomplete from a geometrical one: non-manifold topology, missing or unclear surface normals, inadequate wall thickness, dimensional inaccuracy, and surface intersection artefacts—which are undetectable in rendered imagery but disastrous in manufacturing contexts—all present.

For example, at the object scale in Chap. 2, AI-generated product forms as mesh geometries need consistent human mediation and post-processing before being manufactured. At the architectural scale of Chap. 3, the distance between AI facades and BIM-compatible solid geometry (required from structural analysis, environmental assessment, regulatory submission, etc.) was documented. In Chap. 4, dealing with the urban scale, it was demonstrated by the discrepancy between figure-ground configurations produced by AI and the georeferenced, dimensionally precise urban models needed for development and planning permission. Finally, at the platform level in Chap. 5, it served to motivate the need for specific CAD-native generative models, fabrication-aware topology optimization, and agentic design-to-CAE pipelines that represents a series of attempts to move AI generation closer to the required real-word conventions of manufacturing.

6.2.2 *The Gap's Operative Manifestations*

Concerning manufacturing processes and practices, there is a recurring set of geometric conditions through which the AI-to-manufacture gap manifests itself, involving AI tools, object types, and fabrication processes. According to Ngo et al. [7], some primary categories in additive manufacturing geometry preparation are non-manifold edges (two faces shared more than edge: a topological ambiguity that no slicer can solve); missing or inverted surface normal; self-intersecting surfaces (object faces that pass one through another); insufficient wall thickness (surfaces that are below the minimum thickness and that generated printed walls that cannot be deposited at the specified layer height); and dimensional inaccuracy (scale factors that are far from real-world dimensions).

Each AI tool category has a different relative frequency and impact concerning these issues. The gap is introduced at the transition from image to mesh reconstruction by text-to-image diffusion models, which do not generate three-dimensional geometry. Topological error is primarily introduced during the image-to-3D conversion step (whether via Zero-1-to-3 [8], TRIPO™ AI, or similar systems).

Text-to-3D and image-to-3D systems, even if they can directly generate three-dimensional geometries, lack in terms of topological precision: almost all their outputs show the non-manifold and self-intersection conditions documented above, because their training objectives optimize for visual fidelity rather than topological correctness. Domain-fine-tuned systems, where the LoRA-based rim design pipeline introduced by Li et al. [3] is located, can not only reduce these issues but also completely eliminate them. Practically, such systems try to narrow the training distribution to domain-specific geometries, with the aim of improving the statistical possibilities that mechanically coherent outputs could be generated.

6.3 The Five-Phase AI-To-Fabrication Design Pipeline

As an original methodological contribution, this chapter proposed the Five-Phase AI-to-Fabrication Design Pipeline: an attempt to organize the design-to-manufacture process on five consecutive and connected phases, where each of them is related to a specific tool, human mediation activity, and clear transition criteria from a phase to the other.

The framework is not dedicated to a specific AI tool because, for example, as the panorama of generative AI systems is under continuous evolution, Phase 2 can be deployed for different tools without structural modification in its theoretical foundations. The cyclical nature of design development is accommodated via an iteration loop between Phases 2 and 4, allowing the pipeline to be traversed again as geometric and fabrication feedback from later phases informs the revision of previous decisions. A thorough explanation of each stage is provided below, based on the information gathered in the earlier chapters and the larger body of the literature.

6.3.1 Phase 1—Brief Definition

Operative Logic: Phase 1 is entirely analogue and does not include any AI tools since it has the main role of defining the constraints that all the following AI-assisted activities must follow: the fabrication process parameters that limit achievable geometry (minimum wall thickness, maximum overhang angle, support structure constraints, dimensional accuracy tolerance); the material properties determined by the fabrication process and use context; the dimensional requirements derived from ergonomic and functional specifications; and the functional zone specification, which explicitly identifies geometrically or functionally critical regions of the design that must be shielded from AI modification throughout the pipeline.

Furthermore, the functional zone specification is of particular importance, and it was presented as a methodological contribution from Faruqi et al. [2] through Style2Fab™. The researchers show that applying an uncontrolled amount of AI stylization to fabricated objects can structurally modify some critical regions—mating surfaces, closure geometries, load-bearing interfaces—thus impacting the integrity of fabrication. Considering this, the constraints identified in Phase 1 create some fundamental design guidelines that are transmitted progressively to Phases 2, 3, and

4, and ensure that AI generation and geometric reconstruction are confined between well-defined boundaries that can preserve objects' functionality. Phase 1 does not work only at the object scale but also at the other ones presented: at the architectural one, it defines the building's regulatory constraints, programmatic requirements, and structural system specifications; at the urban scale, it establishes planning regulations, ecological performance targets, and community participation requirements.

6.3.2 Phase 2—AI-Assisted Ideation

Operative Logic: In Phase 2, AI generative tools are used to create design alternatives that must be coherent with the set of constraints identified in Phase 1. The methodology proposes the combination of text-to-image diffusion models (Stable Diffusion™ XL [7], Midjourney™, DALL·E 3™ [4]) with image-conditioning models (ControlNet [9]) to generate aesthetically different outputs but coherent from a formal perspective. Through their Sketch2Prototype™ study, Edwards et al. [1] illustrated that the implementation of a text-based description/brief between sketch input and image generation can contribute to more diverse outputs and reduce design fixation, the psychological condition where early representation limits further explorations [10]. A comprehensive evaluation methodology including formal coherence, structural rationality, fabrication constraint alignment, and design innovation is used to score the produced alternatives against the Phase 1 design brief.

Among the different phases, Phase 2 is the only one more directly impacted by the evolution of the current AI tools landscape. For example, architecture-specific platforms like FORMAS.AI—through its Style Memory system and multi-view consistency—address the issue of the coherence between outputs that characterizes many of the general-purpose diffusion tools, while the domain-inappropriateness issue of generic diffusion models in engineering contexts is addressed by domain-fine-tuned models, such as Li et al.'s LoRA-based solution [3]. The structured prompt templates and evaluation rubrics of Phase 2 will not change as these platform capabilities develop, but the particular AI systems that fill them will. This is a deliberate design choice in the framework's architecture that will have a significant impact on its longevity as a methodological tool.

6.3.3 Phase 3—Geometric Reconstruction

Operative Logic: Phase 3 is the one where the AI-to-manufacture gap happens more consistently. The selected outputs from Phase 2 are translated from bidimensional images into watertight, fabrication-ready three-dimensional geometry using a combination of parametric CAD reconstruction, image-to-3D conversion, and mesh validation. The original AI-generated images are used as a design intent and a formal reference rather than a direct import, because, as highlighted in existing literature

[1–3, 11], the lack of geometric or material meta-data makes the images not directly transferable into fabrication-ready objects without the intervention of human mediation. At the expense of the most reconstruction work, the parametric reconstruction pathway—using Rhinoceros3D ™ with Grasshopper ™, Blender ™, or comparable parametric modelling environments—offers the best degree of geometric control and fabrication compliance. The image-to-3D conversion method, which uses TRIPO ™ AI, Shap-E [5], or similar systems, provides speed and accessibility at the expense of geometric dependability. Prior to every production effort, the structured mesh validation sequence outlined below must be completed. Instead of a simple visual inspection, a series of quantitative checks are used in the validation phase: among them are wall thickness analysis scripts (the minimum accepted threshold from fabrication process); mesh integrity verification (to verify the absence of non-manifold edges, surface normal coherence, self-intersection identification, etc.); the dimensional accuracy specified in Phase 1 brief; and the functional zone integrity confirmation. The failure of one of these assessment criteria bring the design loop back to Phase 2 or to parametric corrections within Phase 3, with a maximum of three failures before the design direction is considered non-viable thus to be abandoned.

6.3.4 Phase 4—Prototyping and Iteration

Operative Logic: In Phase 4, the selected and assessed mesh geometry is transferred into a physical form through fabrication process. The objective of the produced object is two-fold: it represents a physical realization of the specific design intent discussed in Phase 1—and visualized in Phase 2—and it is a manufacturing validation that can detect geometric discrepancies not visible in the digital mesh—e.g., thermal warping, layer adhesion failure, support structure interference with functional geometry—and that requires corrections in Phase 3 before any manufacturing attempt. Phase 3 parametric correction and Phase 4 re-fabrication are triggered when physical measurements at selected control points deviate from the Phase 1 dimensional specification by more than the specified tolerance.

As theorized in the design prototyping literature by Camburn et al. [12], the framework explicitly distinguishes between the looks-like prototype, which is the AI-generated imagery that conveys design intent and permits early-stage client or team evaluation, and the works-like prototype, which is the physically manufactured object that demonstrates functional and dimensional adequacy.

Such a distinction is not only semantic: it confirms how AI generation and physical prototyping are delegated to different tools and at a different level of geometric consistency and precision, covering different purposes and objective within a design process. Indeed, if the looks-like prototype, derived from Phase 2, is validated in terms of formal novelty, aesthetic coherence, and design intent, the works-like prototype—the result of Phase 4—is analysed in terms of dimensional accuracy, structural integrity, and functional performance. The merging of these methods—therefore considering AI-generated images as a direct evidence of fabrication readiness—is the main source of prototyping failures that actually characterize many of the existing naive AI-assisted design-to-fabrication practice [11].

6.3.5 Phase 5—Test and Evaluation

Operative Logic: Phase 5 represents an analytic evaluation of the produced physical prototype in relation to the original design brief from Phase 1, and uses categories that cover geometric fidelity (dimensional accuracy at all control points), surface quality (layer height compliance, surface roughness within specified bounds), structural integrity (absence of delamination, warping, or stress concentrations), and functional performance (assessment of the object's capacity to fulfil its intended use within the constraints specified in Phase 1). The expected outcomes deriving from this evaluation process are three: full compliance—where the prototype fulfils all the criteria and the pipeline can be considered concluded; partial compliance—a return to Phase 3 for further corrections is required; and non-compliance—the design direction is inadequate and requires a return to Phase 2 where the design intent needs to be revised again in relation to the Phase 1 brief.

Using text as an intermediary modality between sketch and generation [1], encoding functional zones before AI engagement [2], applying domain fine-tuning where domain-specific training data is available [3], validating geometry quantitatively before 3D conversion [11], and maintaining explicit separation between looks-like and works-like prototyping [12] are five best practices that are synthesized from the cross-study AI-to-fabrication literature and from the experimental evidence generated within this volume. These five phases structure the framework, providing operative guidance at each stage. This proposed framework aims also to be a teachable methodological protocol for design education and design curricula, while also addressing some problems identified in professional practice to effectively assess and solve the existing AI-to-manufacture gap at a multiscale level.

6.4 Experimental Case Study: AI-Generated Backpack from Ideation to Fabrication

A backpack model is the authors' experimental case studies presented in this chapter and that represents a direct application of the Five-Phase framework to the design and physical fabrication. The typology of a backpack has been chosen because it is a consumer object that combines organic surface complexity, specific pocket geometry, and some functional requirements that are linked to human body dimension. Using this case study, the full methodological pipeline is document: from design brief to mesh validation; from dimensional corrections to slicer configuration; and final manufacturing process. In each step, the specific geometric conditions characterizing each boundary conditions required human mediation. Furthermore, the backpack presents a series of multiple curved surfaces that are extremely challenging for image-to-3D reconstruction systems, since many problematic artefacts concentrate themselves at surface intersection regions, indeed creating a stress moment especially for Phase 3 and in relation to the ergonomic constraints framework of Phase 2 [11].

6.4.1 Phase 1—Design Brief and Constraint Specification

The design brief was created in accordance with the Five-Phase framework's Phase 1 methodology before any AI tool engagement. With a height-to-width ratio based on ergonomic data for single-compartment backpack form, the dimensional specification set a goal bounding box of roughly 138 × 135 × 200 mm, which is commensurate with a compact daypack dimensioned for a young adult user. The Creality™ K1 Max FFF process parameters, which were consistently used in this research program, were used to set fabrication constraints. These included a minimum wall thickness of 0.48 mm (two wall loops at 0.24 mm layer height), a maximum self-supporting overhang angle of 45°, and a requirement for flat-bottom contact geometry for heated bed adhesion without a brim.

Regarding the functional zone specification—related to the locked-zone principle introduced by Faruqi et al. [2] in their Style2Fab system—it identified three critical geometric areas: the strap attachment zones (should strap and top carry handle); the zip track interface which is the main compartment close; and the base geometry for bed adhesion. These three areas were identified as needing geometric integrity preservation via Phase 4 fabrication and Phase 3 rebuilding, and their Phase 1 specification offered the clear standards by which the Phase 5 functional evaluation would be carried out.

6.4.2 Phase 2—AI-Assisted Ideation and Concept Generation

In Phase 2, the combination of text-to-image diffusion model generation and image conditioning was used to produce a set of design alternatives with the same formal and dimensional constraints identified in Phase 1. The text prompts were structured towards the generation of a compact everyday backpack with clean geometry, muted material palette, and a minimal presence of pockets. From an initial population of alternatives (Fig. 6.1), the selection was reduced to the one that best incorporated the design constraints specific to Phase 1. Before moving on to Phase 3 reconstruction, the chosen concept direction—which has a rounded main body with a single front pocket and integrated strap geometry—further improved using image-to-image conditioning to guarantee consistent multi-view coherence. The particular criteria that would be

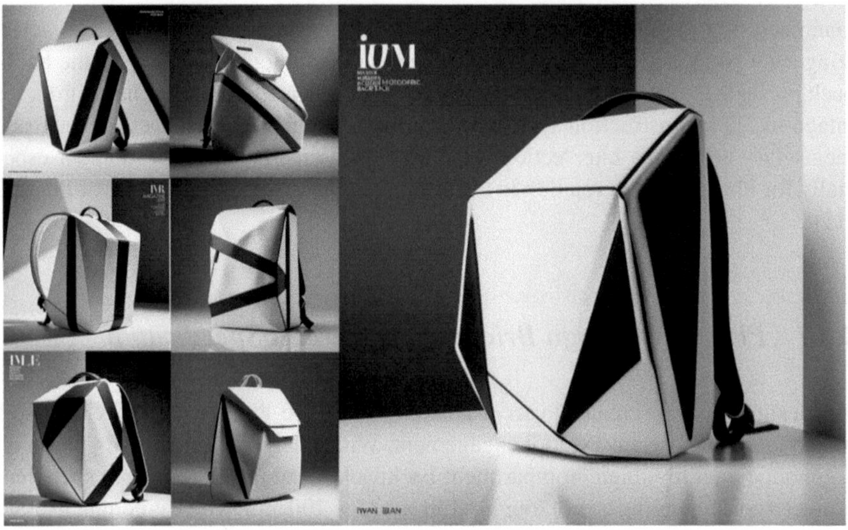

Fig. 6.1 The set of generated images and the selected one

used for the Phase 5 functional evaluation were provided by Phase 4 fabrication and their specification in Phase 1.

A crucial decision made in Phase 2, and impacting also Phase 3, was the choice to be made between a parametric CAD reconstruction or an image-to-3D conversion to generate a manufacturable outcome. Due to the compound surface of the backpack—and its multiple curved intersection in several delicate areas—the parametric pathway was considered too effort-intensive and an image-to-3D conversion approached was selected. TRIPO™ AI, a commercial platform for geometric reconstruction starting from a bidimensional reference image, was selected with the expectation that the generated output would anyway require structural validation and human mediation through Phase 3 before being suitable for fabrication phase.

6.4.3 Phase 3—Geometric Reconstruction and Mesh Validation

500,473 faces, 750,708 edges, and 250,223 vertices composed the image-to-3D reconstruction output generated from TRIPO™ AI (Fig. 6.2). The high number is a direction reflection of the backpack complex geometry, characterized by curved strap attachment surface and pocket intersection along the main object body. The produced mesh was then exported in STL format to start the validation sequence related to Phase 3.

There were no significant topological defects found in the first topological validation, which included non-manifold edge identification, surface normal coherence

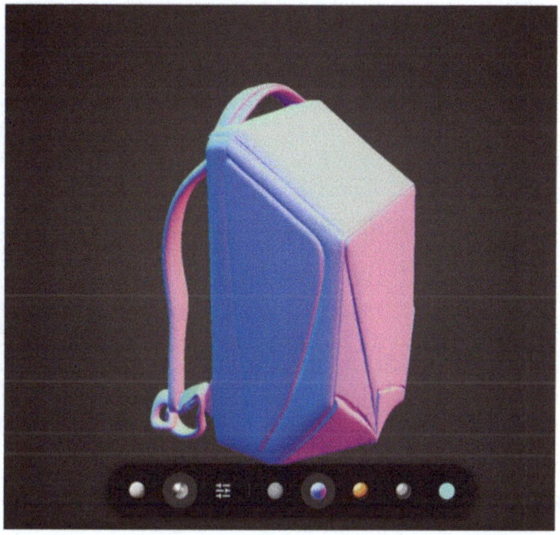

Fig. 6.2 The reconstructed model using TRIPO™ AI

check, and self-intersection analysis, and confirmed the effectiveness of the image-to-3D reconstruction model to avoid all the most common topological failures. Nevertheless, the absence of such problems that does not automatically make the generated output ready for fabrication and, furthermore, two important geometric conditions need to be corrected before proceeding with the slicing phase.

First, design inaccuracy: to reach the target dimensions of 138.20 × 134.61 × 200.00 mm specified in the Phase 1 brief, the mesh bounding box as reconstructed (1,382 × 1,346 × 2,000 units) needed to be rescaled by a factor of 0.1. Such rescaling operation is related to a well-known limitation of current image-to-3D reconstruction systems since they generate geometries that are positioned in an arbitrary tridimensional space without relation to any real-world convention thus requiring a crucial scale correction phase before any manufacturing attempt [7]. The second condition was geometric orientation, since the produced meshed sat on a curved base surface that would create issues in terms of bed adhesion during fabrication phase. To address this issue, the model was manually rotated and to maximize the contact surface with the heated print bed, following a common procedure in FFF workflows and in line with Phase 1 constraint specifications related to flat-bottom contact geometry (Fig. 6.3).

Examining the mesh geometry at the validated surface junctions (base contact surface, pocket closure geometry, and strap attachment zones), it was discovered that there were no double-surface or overlapping surface conditions at any of the three functional zones listed in Phase 1.

In conclusion, the Phase 1 functional zone specification directed the efforts in Phase and also demonstrated the efficiency of the locked-zone methodology: indeed, through the identification of some critical zones before the AI implementation phase, the needed Phase 3 validation was oriented directly towards the most geometrical

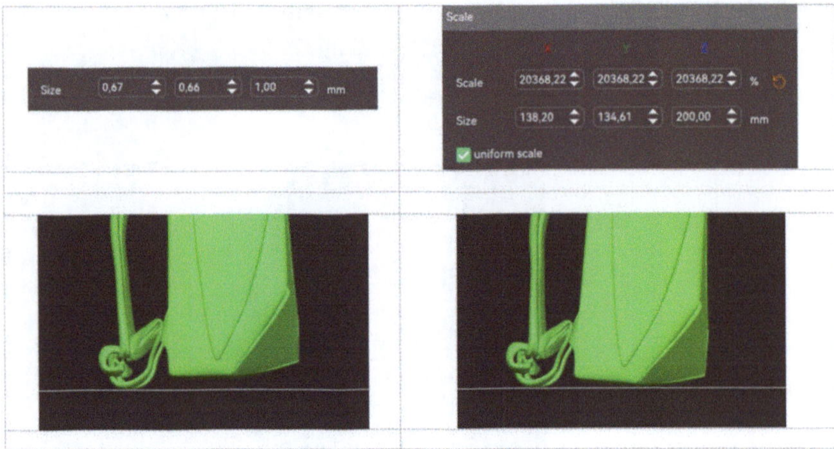

Fig. 6.3 The rescaling and rotation operation

problematic regions of the backpack geometry and was not treated as mere generic mesh health check.

6.4.4 *Phase 4—Fabrication and Process Documentation*

After Phase 3 (validation), the oriented and corrected mesh was sent to the slicer environment set up for the Creality™ K1 Max FFF printer with NEEMA3D™ PLA:EVO filament. All of the slicer configuration was completed in approximately 14 min. A required tree support was created by the slicer to solve the overhanging geometry at the strap roots and the undercut of the front pocket: a correction coherent with the Phase 1 specification of a needed support strategy through automatic support insertion (Fig. 6.4 and Table 6.1).

The backpack model's physical production continued uninterrupted. After cooling, the finished print—which still had the tree support attached—was taken out of the heated bed and put through post-processing cleaning and surface correction (17 min) as well as support removal (12 min). Table 6.2 records the entire process time from slicer setting to post-processed prototype.

In total, the fabrication needed approximately 3 h and 39 min, showing the geometric complexity of the model (500,473 faces) and high needed quantity of material deposition volume. Furthermore, the organic surface geometry characterizing the backpack generated a more complex toolpath than what happens with a more common and simple rotational symmetry of container objects and, for this reason, required a longer deposition time with a print speed of 100 mm/s. A full pipeline traversal from concept to fabricated prototype for a complex consumer object takes

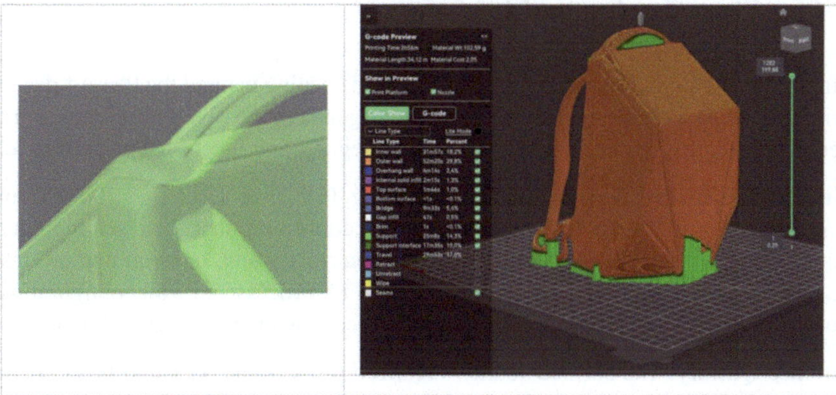

Fig. 6.4 The support insertion and the slicer environment

Table 6.1 Creality™ K1 Max FFF printer's print parameters

Print parameter	Value
Layer height	0.24 mm
Nozzle temperature	230 °C
Bed temperature	50 °C
Wall loops	2
Top/bottom shell layers	5
Infill density	15%—Grid pattern
Print speed	100 mm/s
Support type	Automatic tree support
Material	NEEMA3D™ PLA:EVO, 1.75 mm

Table 6.2 Process time from slicer setting to final prototype

Process phase	Duration
Slicer configuration	14 min
Print time	2 h 56 min
Support removal	12 min
Cleaning and final corrections	17 min
Total process time	~3 h 39 min

about half a working day of total elapsed time, which is consistent with the framework's deployment in both studio teaching sessions and professional prototyping workflows [12]. This result provides a useful benchmark for the design iteration economics of the Five-Phase framework.

6.4.5 Phase 5—Test and Evaluation

Finally, the manufactured prototype was assessed keeping in mind the Phase 1 design brief and the four dimensions in the evaluation agenda of the Five-Phase framework: geometric fidelity, surface quality, structural integrity, and functional performance.

Geometric Fidelity: The prototype, measuring 138.20 × 134.61 × 200.00 mm, was coherent with Phase 1 specification, while confirming that the rescaling operation conducted in Phase 3 had been realized correctly and, more important, that no dimensional distortion was identified in the FFF process and wall configuration. Such results testify a full compliance with the needed geometric fidelity dimensions.

Surface Quality: The surface quality—realized through the 0.24 mm layer height and two-wall-loop configuration—was considered satisfying even though the curved surfaces of the main body and front pocket showcased some of the FFF layer stepping artefacts. Furthermore, there were no surface irregularities or delamination that could have been caused by slicer or process error. At the given process settings, surface quality is evaluated as compliant for a working prototype.

Structural Integrity: Considering that the backpack was a non-load-bearing prototype, its body showed an adequate structural integrity at the 15% grid infill configuration. Concerning the strap attachment zones—detected as critical regions in Phase 1 and validated in Phase 3—those were found physically intact in the printed prototype without the presence of layer separation phenomena. Such a positive outcome validates the methodology based on functional zone specification in Phase 1 and the validation process of Phase 3: the fabrication outcome at these crucial areas was successful because the strap shape was recognized and safeguarded throughout the pipeline.

Functional Performance: All the main compartments and pocket closure were assessed as physically coherent and structurally continuous in the produced prototype. Furthermore, the absence of geometric artefacts—already confirmed in Phase 3 and now validated—led to a successful functional outcome at Phase 5: indeed, as intended by the Phase 1 functional zone specification, the pipeline's focused validation effort had eliminated the principal geometric risk prior to fabrication commitment (Table 6.3).

Table 6.3 Geometric properties of the backpack mesh as assessed in Phase 3

Model property	Value
Vertices	250,223
Edges	750,708
Faces	500,473
File size	24.438 KB
Final dimensions (after correction)	138.20 × 134.61 × 200.00 mm

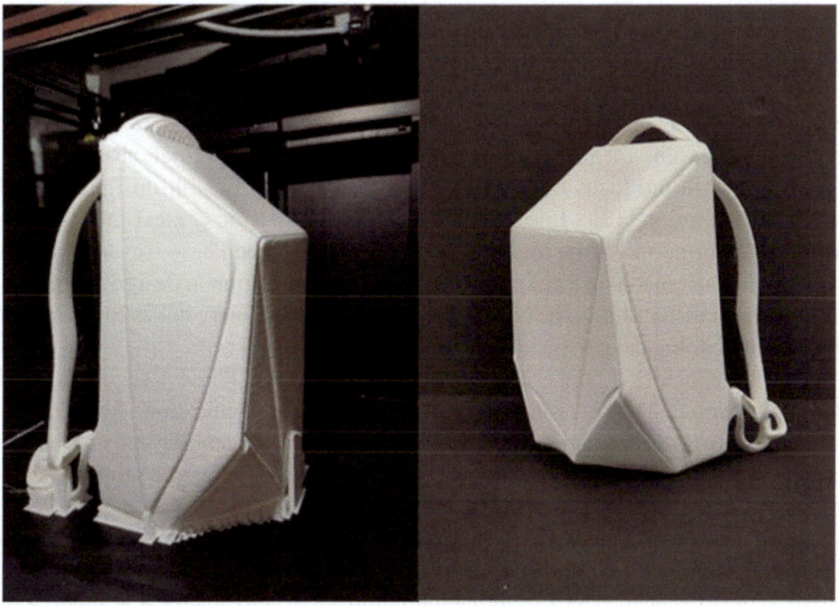

Fig. 6.5 The printed sample with and without the supports

In the end, the Phase 5 evaluation outcome for the backpack case study prototype can be considered in compliance with the needed requirements: the geometric fidelity, surface quality, structural integrity, and functional performance criteria established in Phase 1 were satisfied by the prototype in a coherent tolerance rate with a first-iteration FFF concept prototype (Fig. 6.5).

Such as positive outcome was possible also thanks to the functional zone specification of Phase 1—e.g., the identification of the strap attachment points, geometry of the pocket, and the base contact issue with the heating bed—and the validation effort in Phase 3. The implementation of the functional zone constraints throughout the methodology pipeline allowed Phase 3 validation to focus its effort towards the regions subjected to the highest geometrical risk in order to avoid topological problems and possible failures before the manufacturing phase.

6.5 Framework Validation: Synthesis of Findings

When compared to the larger literature reviewed in Sect. 6.2, the experimental data produced by the backpack case study supports a number of methodologically and practically significant implications for the Five-Phase framework and for AI-assisted design-to-fabrication practice in general.

The first implication is related to the connection between Phase 1 brief and Phase 5 outcome. The precise identification of the functional zone specification—as already mentioned, the strap attachment points, pocket closure geometry, and base contact surface as critical regions requiring directed validation in Phase 3—was one of the primary reasons for the manufacturing success. This confirmation is coherent and in line with the focus that design literature currently exhibit on front-loaded constraint definition [12] while extending also to the specific context of AI-driven workflows. Furthermore, the importance given to an AI-free Phase—even though it might appear debatable in a chapter focusing on AI integration in design and manufacturing processes—has to be considered to only precautionary but as fundamental from a methodological perspective.

The first conclusion concerns the relationship between Phase 1 specificity and Phase 5 outcome. The precision of the functional zone specification—the identification of strap attachment points, pocket closure geometry, and base contact surface as critical regions requiring directed validation in Phase 3—was the primary determinant of fabrication success. This finding is consistent with the design prototyping literature's emphasis on front-loaded constraint definition [12], and extends it to the specific context of AI-assisted workflows: the greater the ambiguity tolerated at Phase 1, the greater the risk of fabrication failures that only manifest at Phase 4 or Phase 5, when correction costs are highest. The framework's insistence on a fully analogue, AI-free Phase 1—which may appear counterintuitive in a chapter advocating AI integration—is therefore methodologically essential rather than merely precautionary.

The second conclusion concerns the universality of the dimensional correction requirement. When reconstructed, the bounding box of the backpack mesh measured 1,382 × 1,346 × 2,000 units, and required a scaling factor of 0.1 to reconcile with the specifications in Phase 1 brief. This reported dimensional mismatch confirms the overall behaviour document in AI-to-manufacture literature [11] for image-to-3D reconstruction systems towards multiple object categories and platforms. More specifically, such systems tend to generate geometries in an arbitrary space without any connection to real-word dimension conventions, and this is due to the fact that their training material is more image-based rather than structured upon precise engineering drawings. Indeed, since the Phase three-dimensional correction phase is a structural feature of current image-to-3D technology rather than an object-specific artefact, it must be seen as a required pipeline element rather than an optional corrective.

Finally, the third implication regards the scalability of this framework as a professional and educational tool. The measured time from slicer configurations to the final physical prototype was approximately 3 h and 39 min in the framework of a normal half-day design session. This result confirms that the Five-Phase framework is reasonable under a time perspective for both design education (a design studio session) and professional work (a rapid iteration cycle). In both contexts, the framework's value has not to be found in a reduced fabrication time—the FFF print time is the result of the machine process and geometric parameters and it is not influenced by any specific workflows—but, undoubtedly, in the currently trial-and-error sequence that characterizes the majority of the existing AI-assisting prototyping practices that can be substituted by a more predictable sequence of human-AI cooperation activities.

6.6 Prospective Developments: Towards a Closed Design-To-Fabrication Loop

The aim of the Five-Phase framework presented in this chapter is to offer a contribution to close the AI-to-manufacture gap through a structure human mediation, intended as the clear identification of phase boundaries, validation standards, and iteration procedures that guide the designer's experience and judgement to the most important stages of the pipeline. Nevertheless, this very human-based approach is at the same time the primary strength of the framework but also its primary limitation since it requires the presence of a designer with a satisfying theoretical and technical knowledge to proficiently identify Phase 1 constraints, evaluate alternatives in Phase 2, validate outcomes in Phase 3, and finally assess the manufactured outcome in Phase 5 in relation to the original design brief. Undoubtedly, the documented technological developments in Chap. 5 will have sensible improvements in the follow years, generating also a consistent shift in the balance of human and automated mediation documented into this section. In this perspective, three are the developments of particular significance.

First is the maturation of CAD-native generative models that can produce parametric construction sequences and outcomes that can be directly imported in engineering CAD environments, and eliminating the mesh-to-BRep conversion challenge that represents the most challenging issue in Phase 3. According to Wu et al. [13], autoregressive transform models have the ability to produce CAD construction sequences that, if executed, can generate already valid and editable solid geometries. Therefore, as these models mature there will be a sensible reduction in the reconstruction effort analysed in Phase 3.

Second is the rise of fabrication-aware generative model that could directly implement manufacturing constraints (e.g., minimum wall thickness, support structure requirements, build orientation considerations) as an explicit training objective, and generate outputs closer to a final fabrication-ready form. The principle is demonstrated by the topology optimization techniques discussed in Chap. 2, which impose manufacturing process constraints as hard conditions within the optimization. Its extension to the organic, aesthetically driven generative contexts of Phases 2 and 3 is an active research frontier [14] with direct implications for the pipeline's present reliance on Phase 3 manual validation [15, 16].

Third is the development in agentic design pipeline already documented in Engineering.ai [17], in regard to specialist AI agents—each equipped with domain knowledge and simulation tool access—coordinated among them and operating under a chief engineer agent to autonomously create simulation workflow. In relation to the Phase framework, such multi-agent architecture would delegate Phase 1 brief and constraint definition to a brief-parsing language agent; Phase 2 to a generation agent; Phase 3 to a mesh analysis and correction agent; Phase 4 to a manufacture control agent; and Phase 5 to an assessment agent, with a human designer only involved in defining the design critical design point that influence all the generative and manufacturing processes. Furthermore, the Five-Phase framework will not be eliminated by the concretization of this vision but rather would be automated while preserving human-AI design collaboration.

6.7 Conclusion

This chapter served as a platform to introduce the Five-Phase AI-to-Fabrication Design Pipeline that has a methodological and operative synthesis of this volume in the form of an original framework that could address the AI-to-manufacture gap that is considered as one of the main bottlenecks in the implementation of AI-assisted design in professional and educational practices. The five phases—Brief Definition, AI-Assisted Ideation, Geometric Reconstruction, Prototyping and Iteration, and Test and Evaluation—have ben theoretically refined; situated in the multiscale theoretical and technical context of this book; and then validated through an authors' original case study attempt.

Furthermore, the backpack examples showed of the Five-Phase framework are efficiently deployable within a studio or workshop timeframe (approximately 3 h and 39 min for the whole process), and the functional zone specification in Phase 1 is fundamental for the fabrication success of Phase 5. The empirical evidence validates the fundamental methodological claim of the framework: that the AI-to-manufacture gap is a structured set of geometric conditions (dimensional inaccuracy, topological artefacts, surface normal ambiguity, insufficient wall thickness) that a rigorous, explicitly phased pipeline can systematically identify, locate, and address before physical fabrication is attempted.

The five operative best practices presented in the framework—the use of text as an intermediary modality between sketch and generation; the definition of functional zones before AI usage; the application of fine-tuning domain where data is available; the qualitative validation of the geometries before a 3D conversion; the distinct separation between two-dimensional imagery and three-dimensional real-world prototyping—represent a transmittable and teachable protocol for the integration between AI-generative models and design-to-fabrication practices.

Moreover, the relevance of these five best practices spans through all the scale hierarchy presented in the volume: from the object scale, to the architectural and urban scale one. Even though in each of these contests the AI tools, fabrication processes, and constraint specifications of Phases 1 and 2 will be different, the pipeline organizing logic will be remain the same.

The future developments described in Sect. 6.6—CAD-native generative models, fabrication-aware generation architectures, and agentic multi-agent design pipelines—will definitely automate many of the operative points of the framework but not substitute its structural logic. Therefore, if this research field is moving more and more towards autonomous design-to-fabrication pipelines as a long-term aspiration, the Five-Phase's explicit phase structure, transition criteria, and human judgement integration points represent a proper template to incrementally deploy AI automation considering that each step forward in AI capability might reduce the human effort without changing the framework's inner logic The framework's most enduring contribution to the process of AI-assisted design and prototyping is its adaptability, or its ability to incorporate new AI capabilities while maintaining the organizing logic that makes the pipeline predictable, auditable, and educationally transferable.

References

1. Edwards KM, Man B, Ahmed F (2024) Sketch2Prototype: rapid conceptual design exploration and prototyping with generative AI. Proc Des Soc 4:1989–1998. https://doi.org/10.1017/pds.2024.201
2. Faruqi F, Katary A, Hasic T et al (2023) Style2Fab: Functionality-aware segmentation for fabricating personalized 3D models with generative AI. In: Proceedings of the 36th annual ACM symposium on user interface software and technology (UIST '23). ACM, New York, pp 1–13. https://doi.org/10.1145/3586183.3606723
3. Li K-Y, Huang C-K, Chen Q-W, Zhang H-C, Tang T-T (2025) Generative AI and CAD automation for diverse and novel mechanical component designs under data constraints. Discov Appl Sci. https://doi.org/10.1007/s42452-025-06833-5
4. Betker J, Goh G, Jing L et al (2023) Improving image generation with better captions. OpenAI Technical Report
5. Jun H, Nichol A (2023) Shap-E: generating conditional 3D implicit functions. arXiv:2305.02463
6. Zeng X, Vahdat A, Williams F et al (2022) LION: Latent point diffusion models for 3D shape generation. In: Advances in neural information processing systems (NeurIPS), New Orleans
7. Podell D, English Z, Lacey K et al (2023) SDXL: improving latent diffusion models for high-resolution image synthesis. arXiv:2307.01952
8. Liu R, Wu R, Van Hoorick B, Tokmakov P, Zakharov S, Vondrick C (2023) Zero-1-to-3: zero-shot one image to 3D object. In: Proceedings of the IEEE/CVF international conference on computer vision (ICCV), pp 9298–9309. https://doi.org/10.1109/ICCV51070.2023.00853
9. Zhang L, Agrawala M (2023) Adding conditional control to text-to-image diffusion models. In: Proceedings of the IEEE/CVF international conference on computer vision (ICCV), pp 3836–3847. https://doi.org/10.1109/ICCV51070.2023.00355
10. Jansson DG, Smith SM (1991) Design fixation. Des Stud 12(1):3–11. https://doi.org/10.1016/0142-694X(91)90003-F
11. Ngo TD, Kashani A, Imbalzano G, Nguyen KTQ, Hui D (2018) Additive manufacturing (3D printing): a review of materials, methods, applications and challenges. Composites Part B Eng 143:172–196. https://doi.org/10.1016/j.compositesb.2018.02.012
12. Camburn B, Viswanathan V, Linsey J et al (2017) Design prototyping methods: state of the art in strategies, techniques, and guidelines. Des Sci 3:e13. https://doi.org/10.1017/dsj.2017.10
13. Wu R, Xiao C, Zheng C (2021) DeepCAD: A deep generative network for computer-aided design models. ICCV 2021, pp 6772–6782
14. Sigmund O, Maute K (2013) Topology optimization approaches: a comparative review. Struct Multidiscip Optim 48(6):1031–1055. https://doi.org/10.1007/s00158-013-0978-6
15. Perna V, Kyratsis P (2025) Diffusion models in product design: optimizing form toward digital prototyping. CAD/CAM: Comput Aided Design Manufa 17:21

16. Perna V (2025) Technicity between prediction and autonomy: Ceci n'est pas une chaise (This is not a chair). In: Digital product design and manufacturing, pp 179–197. CRC Press
17. Xu R, Qi Y, Feng J, Chu X (2025) Engineering.ai: A platform for teams of AI engineers in computational design. arXiv:2511.00122. https://arxiv.org/abs/2511.00122

MIX
Papier aus verantwortungsvollen Quellen
Paper from responsible sources
FSC® C105338
FSC
www.fsc.org

If you have any concerns about our products,
you can contact us on
ProductSafety@springernature.com

In case Publisher is established outside the EU,
the EU authorized representative is:
Springer Nature Customer Service Center GmbH
Europaplatz 3, 69115 Heidelberg, Germany

Printed by Libri Plureos GmbH
in Hamburg, Germany